# THE DOVECOTES OF SUFFOLK

by

John McCann

with a Foreword by

Professor R. W. Brunskill, OBE, FSA

and an Appendix by

Dr Rosemary Hoppitt

SUFFOLK INSTITUTE OF ARCHAEOLOGY AND HISTORY
1998

Published 1998 by

The Suffolk Institute of Archaeology and History
Oak Tree Farm, Hitcham, Ipswich, Suffolk IP7 7LS

ISBN 0 9521390 1 4

© John McCann, 1998

All rights reserved. No part of this publication may be reproduced or transmitted in any form or by any means, electronic or mechanical, including photocopy, recording, or any information storage or retrieval system, without permission in writing from the publisher.

The publication of this work has been made possible through bequests to the Institute by:

**GWENYTH DYKE** (d. 1992)

**DEREK POWELL** (d. 1994)

and through generous financial help from the following bodies:

**THE COUNCIL FOR BRITISH ARCHAEOLOGY**

**THE D.C. MONCRIEFF CHARITABLE TRUST**

**THE SCOULOUDI FOUNDATION**
in association with
**THE INSTITUTE OF HISTORICAL RESEARCH**

**THE SUFFOLK HISTORIC BUILDINGS GROUP**

Printed in Great Britain by Direct Offset,
27c High Street, Glastonbury, Somerset BA6 9DR

# Foreword

## by Professor R. W. Brunskill, OBE, FSA

Of the many who would put 'enjoying the countryside' as high among their pleasures there can be few who have not been fascinated by dovecotes. Poised socially and physically between the buildings of a privileged household and the buildings of the utilitarian farmyard, their compact shape and their tall and sometimes cylindrical form adds contrast to the land-hugging spread of houses and barns. Externally they are rather mysterious, usually rather bare save for the elaborate louver which serves as beacon, shelter and landing ground. Internally all is clear and straightforward, with patterns deftly illuminated from above of hundreds of nesting boxes, each one providing comfort and privacy.

Recently several books have been published following county and national surveys of dovecotes. Until 1991 they were all prepared against the background of a few rather obscure archaeological studies and a more popular publication of 1920. In 1991, however, John McCann published 'An Historical Enquiry into the Design and Use of Dovecotes'. With a sceptical eye and an enquiring mind he looked for the really reliable evidence for the use of dovecotes, looking in the documents and in the buildings themselves for the uses of dovecotes, for the purposes in erecting them and for the explanations of the various parts which combine to make the buildings so attractive and their study so rewarding. Since 1991 we have all had to look again and more intently at the buildings and to look harder and more sceptically at the received wisdom about those buildings as we have prepared to revise long-held assumptions about them.

John McCann had in mind the study of a single county making use of his new way of looking at dovecotes, less as architectural curiosities and more as functional entities made pleasant to look upon. THE DOVECOTES OF SUFFOLK is the result. Painstaking survey and literary and archival search have been combined to tell the story of these buildings. It is not one with a neat plot or a full set of characters because so many of the dovecotes built in Suffolk have not survived to be studied or appreciated. Nevertheless there is a story to be told and John McCann has told it concisely and authoritatively and all who know and love the subject and the county are grateful to him.

R. W. Brunskill

# Contents

**Foreword by Professor R. W. Brunskill, OBE, FSA** ................................................. 3
List of illustrations ........................................................................................................ 6
Introduction .................................................................................................................. 8
Acknowledgements ...................................................................................................... 10
Glossary ........................................................................................................................ 11
Chapter 1: How dovecotes were used ......................................................................... 14
          What was a dovecote for? ............................................................................ 14
          The production of meat .............................................................................. 14
          The seasonal nature of the supply .............................................................. 14
          Stocking the dovecote ................................................................................. 15
          Feeding ........................................................................................................ 15
          Maintaining the breeding stock .................................................................. 15
          Mineral supplements .................................................................................. 16
          Uses of the manure .................................................................................... 16
          The use of lime-wash ................................................................................. 17
          The scope of this study .............................................................................. 17
Chapter 2: Who could own a dovecote? ..................................................................... 18
Chapter 3: How dovecotes were designed .................................................................. 19
          Pigeons' instinctive responses to birds of prey .......................................... 19
          The perching behaviour of pigeons ........................................................... 19
          Designing the exterior ................................................................................ 19
          Designing the interior ................................................................................ 20
          Protection against birds of prey ................................................................. 20
          Protection against other predators ............................................................ 21
          The introduction of brown rats .................................................................. 21
Chapter 4: The decline of dovecotes ........................................................................... 23
          Trap-shooting ............................................................................................. 23
          The end of the working tradition .............................................................. 24
**The survey of Suffolk dovecotes** ............................................................................. 25
          Dimensions, distances and orientations .................................................... 25
          Statutory Listing ......................................................................................... 25
          A note about ground levels ....................................................................... 25
Chapter 5: Timber-framed dovecotes .......................................................................... 27
          Great Thurlow Hall, Great Thurlow .......................................................... 27
          Badley Hall, Badley ................................................................................... 29
          'South Suffolk A' ....................................................................................... 34
          Church Farm, Clare ................................................................................... 37
          Redcastle Farm, Pakenham ....................................................................... 42
          Chapel Farm, Gipping ............................................................................... 46
          Aspall Hall, Aspall ..................................................................................... 49
Chapter 6: Brick dovecotes .......................................................................................... 53
          Stoke College, Stoke-by-Clare ................................................................... 53
          Flixton Hall, Flixton by Bungay ................................................................ 56
          Goswold Hall, Thrandeston ...................................................................... 60

'South Suffolk B' .................................................................. 61
Crowfield Hall, Crowfield ................................................... 64
Polstead Ponds Farm, Polstead .......................................... 65
Kentwell Hall, Long Melford ............................................... 67
Scotland Place, Stoke-by-Nayland ..................................... 70
The Park, Great Glemham .................................................. 72
Four Bridges, Letheringham ............................................... 76
Hitcham House, Hitcham ................................................... 76
Coldham Hall, Stanningfield .............................................. 78
Cockfield Hall, Yoxford ...................................................... 81

Chapter 7: Stone dovecotes ............................................................... 85
5 Church Walk, Mildenhall ................................................. 85
The Abbey Gardens, Bury St. Edmunds ........................... 87
Marsh Stables, Exning ........................................................ 90
Ousden Hall, Ousden .......................................................... 92
Home Farm, Drinkstone ..................................................... 94
Home Farm, Pettistree ........................................................ 95
Claycotts, East Bergholt ...................................................... 98

Chapter 8: Three oddities ................................................................. 100
Eriswell Hall, Eriswell ....................................................... 100
Alder Carr Farm, Creeting St. Mary ................................ 103
Parsonage Farm, Boxford .................................................. 106

Chapter 9: Comparisons and conclusions ...................................... 108
The sites ............................................................................. 108
Plan forms .......................................................................... 109
Materials ............................................................................. 109
Doorways ........................................................................... 110
Windows ............................................................................. 110
Roofs ................................................................................... 111
Perching ledges ................................................................. 112
Louvers, cupolas and lanterns ......................................... 112
Internal protective devices against birds of prey .......... 113
The provision of nesting places ...................................... 114
The use of unfired clay ..................................................... 114
Internal ledges ................................................................... 115
Potences or revolving ladders ......................................... 115
Other internal features ..................................................... 116
Protection from brown rats .............................................. 116
Conversion to secondary uses ......................................... 117
Distribution ........................................................................ 117
The dating of Suffolk dovecotes ...................................... 118

Chapter 10: What happened to Suffolk's dovecotes? .................. 119
Chapter 11: The future ..................................................................... 121
Notes ................................................................................................... 122
**Appendix by Dr. Rosemary Hoppitt:** Some evidence of Suffolk dovecotes from manorial account rolls ...................................................................... 126

# Illustrations

*(Unless otherwise stated the illustrations are by the author)*

**COLOUR PLATES:**
Front cover: The fifteenth-century pigeon-tower at Stoke College, Stoke-by-Clare, from the north
Back cover: Nest-holes of the ruined dovecote of Goswold Hall, Thrandeston.

*opposite page*

| | | |
|---|---|---|
| 1a | 'South Suffolk A' - interior of roof, with 'pipe' | 42 |
| 1b | Chapel Farm, Gipping - the inner structure of nest-boxes | 42 |
| 2 | The pigeon-tower at Stoke-by-Clare, east elevation | 43 |
| 3a | The dovecote of Flixton Hall, Flixton by Bungay | 58 |
| 3b | The ruined dovecote of Goswold Hall, Thrandeston | 58 |
| 4 | Flixton Hall, interior with potence and feeding platform | 59 |
| 5 | 'South Suffolk B' - interior with incomplete potence | 68 |
| 6a | The derelict dovecote of Crowfield Hall, Crowfield | 69 |
| 6b | The pigeon-tower of Scotland Place, Stoke-by-Nayland | 69 |
| 7 | Kentwell Hall, Long Melford, interior with potence | 72 |
| 8a | The dovecote in Great Glemham park, exterior | 73 |
| 8b | Great Glemham, a round window with decayed wire grill | 73 |
| 9a | The dovecote of Coldham Hall, Stanningfield, exterior | 88 |
| 9b | Coldham Hall, interior with loft and incomplete potence | 88 |
| 10a | Abbey Gardens, Bury St. Edmunds, remains of brick nest-holes | 89 |
| 10b | Home Farm, Drinkstone - the pigeon-tower and farmhouse | 89 |
| 11a | The pigeon-tower of Home Farm, Pettistree | 96 |
| 11b | Claycotts, East Bergholt - the kennels range and pigeon-loft | 96 |
| 12a | The dovecote of Eriswell Hall, Eriswell - exterior | 97 |
| 12b | Eriswell Hall, interior towards the roof | 97 |

*page*

**MAP**
showing the dovecotes of Suffolk, with District boundaries and main towns ...... 27

**FIGURES:**

| | | |
|---|---|---|
| 1 | The timber-framed dovecote of Great Thurlow Hall, exterior | 27 |
| 2 | Great Thurlow, the flight deck and 'pipe' | 29 |
| 3 | The timber-framed dovecote of Badley Hall, Badley, exterior | 30 |
| 4 | Badley Hall, vertical section (by Leigh Alston) | 31 |
| 5 | Badley Hall, isometric drawing (by Leigh Alston) | 32 |
| 6 | The timber-framed dovecote identified as 'South Suffolk A' | 34 |
| 7 | 'South Suffolk A', the interior and nest-boxes | 36 |

## The Dovecotes of Suffolk

| | | |
|---|---|---|
| 8 | The timber-framed dovecote of Church Farm, Clare, exterior | 37 |
| 9 | Church Farm, Clare, isometric drawing (by Leigh Alston) | 38 |
| 10 | Church Farm, Clare, vertical section (by Leigh Alston) | 39 |
| 11 | Church Farm, Clare, cross section (by Leigh Alston) | 40 |
| 12 | Church Farm, Clare, sections of tie-beam and collar | 41 |
| 13 | The timber-framed dovecote of Redcastle Farm, Pakenham | 43 |
| 14 | Redcastle Farm, Pakenham, isometric drawing (by Leigh Alston) | 44 |
| 15 | Redcastle Farm, Pakenham, vertical section (by Leigh Alston) | 45 |
| 16 | Redcastle Farm, Pakenham, isometric detail of gablet (by Leigh Alston) | 45 |
| 17 | The timber-framed dovecote at Chapel Farm, Gipping, exterior | 47 |
| 18 | The dovecote of Aspall Hall, Aspall (photographed in 1947 by H. N. Collinson) | 50 |
| 19 | The pigeon-tower at Stoke College, Stoke-by-Clare, south elevation | 53 |
| 20 | The pigeon-tower at Stoke-by-Clare, interior with nest-holes | 54 |
| 21 | Flixton Hall, interior with doorway and ladder of potence | 58 |
| 22 | The dovecote identified as 'South Suffolk B', exterior | 62 |
| 23 | Polstead Ponds Farm, Polstead, the dovecote and farmhouse | 66 |
| 24 | The dovecote of Kentwell Hall, Long Melford, exterior | 68 |
| 25 | Scotland Place, Stoke-by-Nayland, nest-boxes of clay bats | 71 |
| 26 | Great Glemham, inclined wall-head (by Pamela McCann) | 73 |
| 27 | The dovecote at Four Bridges, Letheringham, exterior | 75 |
| 28 | The pigeon-tower of Hitcham House, Hitcham, exterior | 77 |
| 29 | Coldham Hall, Stanningfield, the window and flight holes | 79 |
| 30 | Coldham Hall, Stanningfield, nest-boxes of clay bats | 80 |
| 31 | The dovecote of Cockfield Hall, Yoxford, exterior (by Tim Buxbaum) | 82 |
| 32 | Cockfield Hall, Yoxford, interior with potence | 84 |
| 33 | The ruins of the medieval dovecote at Mildenhall | 86 |
| 34 | The medieval pigeon-tower in Abbey Gardens, Bury St. Edmunds | 88 |
| 35 | Site plan of St. Edmundsbury Abbey (based on a plan by A. B. Whittingham, by courtesy of English Heritage) | 89 |
| 36 | The dovecote at Marsh Stables, Exning, exterior | 91 |
| 37 | The dovecote of Ousden Hall, Ousden, exterior | 93 |
| 38 | Home Farm, Pettistree, the wrought iron gate from inside | 97 |
| 39 | The dovecote of Eriswell Hall, exterior from north-west | 101 |
| 40 | The windmill buck at Alder Carr Farm, Creeting St. Mary, in 1991 (by Pamela McCann) | 104 |
| 41 | The windmill buck at Alder Carr Farm, as converted in 1996 | 105 |
| 42 | The dovecote of Parsonage Farm, Boxford | 107 |
| 43 | The former dovecote of Henham Hall, Henham (by courtesy of the National Monuments Record Centre, Swindon) | 120 |

# Introduction

The dovecotes of Suffolk have not received the attention they deserve, nor as much as those of neighbouring counties. A major study of Essex dovecotes was published in 1931. There have been two studies of Cambridgeshire dovecotes, one of the whole county in 1977, and a more detailed one of South Cambridgeshire District in 1986. In Norfolk the Norfolk Dovecotes Trust is actively engaged in their restoration. In Suffolk, apart from three pages about dovecotes in a more general article by Claude Messent in 1936, little has been published about them.[1] Why have the dovecotes of Suffolk attracted so little notice?

Like so many issues in the study of dovecotes, it all goes back to Arthur O. Cooke, who first introduced the subject to a wide readership.[2] By the middle of the nineteenth century the traditional dovecotes of Britain had passed out of economic use (for reasons which will be discussed later), and had been forgotten. They remained largely un-noticed by historians and archaeologists until 1887, when R. S. Ferguson collected and published some diverse historical information about them; he followed this with a pioneering study of the dovecotes of Cumberland. His example led others to undertake similar surveys in Herefordshire in 1890 and in Worcestershire in 1905, and then effectively the subject lapsed again.[3]

Interest was revived by Cooke's highly readable 'A Book of Dovecotes', published in 1920. Until then no one had attempted to write a general work on dovecotes on a national scale, for a popular readership. Before this he had written one topographical work, and many books for children on a wide range of subjects, but he had not studied other historic buildings. His book on dovecotes exhibits the merits and defects of that popular approach. He took the historical background from the limited range of material already published. He examined the dovecotes near Edinburgh at first hand, drew on the three earlier county studies, and appealed for information about the dovecotes of other counties in newspapers and magazines. He relied heavily on the replies he received from unknown correspondents, and in many cases he wrote about buildings he had not visited.[4]

For whatever reason, very little information emerged from Suffolk. Under the chapter heading 'Essex and Suffolk' Cooke described ten dovecotes in Essex, and added: 'Suffolk must be passed over with the bare mention of the so-called "dovecote" at Bury St. Edmunds'. More will be said later about what he regarded as an unsatisfactory example. Perhaps he did not manage to place his appeal for information in a journal which reached Suffolk readers - or perhaps they were not sufficiently interested to write to him.

Most later studies of dovecotes have derived the general background from Cooke. He made clear that he never intended his book to be more than an *hors d'oevre*, but it became the only national publication which could be consulted. Cooke's phrases and ideas permeate the whole dovecote literature, have been quoted and re-quoted ever since, and in some respects have misled later writers.[5] If he concluded that there was not much to be said about the dovecotes of Suffolk, apparently most Suffolk people accepted that view.

It is indeed true that Suffolk has retained fewer dovecotes than some other counties. Recent surveys have identified 61 in Herefordshire, 64 in Worcestershire, 70 in Gloucestershire, and well over a hundred in Nottinghamshire.[6] In most counties the number of surviving dovecotes remains unknown. Only 30 have been identified in Suffolk.

On the other hand, those which have survived form an impressive collection. They are of all periods from the fourteenth century to the nineteenth, of all sizes and of almost every plan shape. They are constructed with every building material which is available in the county - timber framing, brickwork, flint rubble and clunch - and two of them have dressings of Barnack stone from Northamptonshire. Unfired clay has been used in several forms for the nest-boxes. They come down to us in every kind of condition, from near-perfect preservation to advanced dereliction. Had there been a survey before the destructive 1950s and 1960s, when so many historic buildings of all kinds were lost, many more would have been recorded. We know from documents, early maps and field-names that dovecotes were formerly as common in Suffolk as everywhere else. They passed out of use owing to major changes in agriculture in the nineteenth century; it seems that in Suffolk those changes affected dovecotes more radically than in some other counties.

This survey was begun in 1991, and has continued intermittently since. The buildings are described as they were when seen. No study of historic buildings can ever be final, but these are all the surviving dovecotes in Suffolk which have been found, after extensive enquiries. Donald Smith found 59 dovecotes in Essex in 1931, but ten more have come to light since, and have been described in the county historical journals.[7] It will not be surprising if more dovecotes are found in Suffolk when their significance is better understood.

## ACKNOWLEDGEMENTS

This study could not have been undertaken without the generous help of several people who are well-informed about Suffolk's historic buildings - Philip Aitkens, Leigh Alston, Mark Barnard, Sylvia Colman, Timothy Easton, and Edward Martin. David Dymond and Peter Northeast kindly answered some historical queries. Others gave valuable help on specific subjects: Pat Ryan helped with the dating of brickwork, Dr. Mark Bailey and Dr. John Birks provided information about polecats, Dr. Graham Twigg about rats, and Peter Dolman about 'the Creeting buck'. Others who have helped are Mrs. P. M. Guild, Ken Robins, Michael Seago, and the staff of Suffolk Record Office. Some illustrations were kindly provided by Leigh Alston, Tim Buxbaum, Pamela McCann, English Heritage and the National Monuments Record Centre, and are credited in the captions. For all of this I am immensely grateful. My wife Pamela accompanied me on some field trips, and was vastly patient while the material was being assembled and written.

We should all be grateful to the owners of these buildings, who generously allowed them to be examined, measured and photographed for this publication. We can best express our gratitude by not intruding on their property without an invitation.


## A NOTE ABOUT WORDS:

Doves are pigeons. In traditional usage a building which housed pigeons was called a 'pigeon house', a 'pigeon-cote', a 'dovehouse', or in some regions a 'culverhouse'. Shakespeare used the word 'dovecote'; and by the eighteenth century 'pigeon-house' and 'dove-cote' were used interchangeably of the same buildings. Since Cooke's work 'dovecote' has become the generally accepted modern term, so it is convenient to use it here. (In Scotland a dove is a 'doo', and the building is correctly called a 'doocot'). In describing a building of two or more storeys in which only the top storey was devoted to housing pigeons the term 'pigeon-tower' is preferable, and will be used more specifically where appropriate. Other terms used here are defined in the Glossary.

# Glossary

*(Words in italics are themselves defined).*

| | |
|---|---|
| Alighting ledge | A projecting ledge outside a tier of *nest-holes* or *nest-boxes*, wide enough for pigeons to perch on, too narrow to provide a refuge for birds of prey. |
| Alighting step | A brick or stone projecting outside a *nest-hole* or *nest-box*, serving the same purpose but not forming a continuous ledge. |
| Assembly marks | Marks inscribed on the components of a timber-framed building during fabrication off site, to identify their positions when assembled on site. |
| Batter | To incline inwards. |
| Bay | The interval between the principal posts of a timber-framed building, or the principal rafters in a roof. |
| Bond | The laying of bricks in a wall according to a regular pattern for strength. |
| Brick nogging | A filling of bricks in the panels of a timber frame. |
| Brick Taxes | Excise duty on bricks and all building products of fired clay, introduced in 1784, which increased the cost by about 15 per cent. Tiles were exempted in 1833; the tax was abolished in 1850. |
| Clamp | A temporary installation in which bricks were fired near the building site, consisting of stacks of *green bricks* interspersed with wood fuel, the whole covered by clay and turf. Combustion was controlled by adjusting the air intakes. |
| Clay bats | Clay with a high chalk content trodden with water and straw to form a plastic material, formed in wooden moulds and air-dried to make shallow rectangular slabs, 1½ to 4 inches thick. |
| Clay daub | Clay trodden with water and straw to make a plastic material which could then be applied like plaster. |
| Clay lumps | A building material composed of clay sub-soil and straw, trodden with water to form a plastic substance, formed in wooden moulds and air-dried to make large rectangular blocks (typically 18 x 9 x 6 inches), much used in East Anglia in the nineteenth century. |
| Closer | Part of a brick which is cut to complete the bond to a vertical edge. |
| Clunch | Chalk which is hard enough to be used for building. |
| Collar | A horizontal timber across two opposite rafters, above the tie-beam and below the apex. |
| Dentils | Rectangular blocks separated by similar spaces, used in a cornice. |
| Dendrochronology | The science by which timbers can be dated by measurement of their growth rings. |
| Dog-tooth course | A course of bricks laid diagonally across the wall so that their corners project and overhang the plain brickwork below, forming a serrated cornice. |
| English bond | A brick *bond* in which alternate courses are composed entirely of *stretchers* or entirely of *headers*. It was common in all brick buildings |

|  |  |
|---|---|
|  | until the seventeenth century, and then was gradually superseded by *Flemish bond* in fashionable contexts. |
| Flared | (of bricks) Over-heated during firing, so that they turn blue or black, often with a vitrified surface. |
| Flemish bond | A brick *bond* in which each course is composed of alternate *headers* and *stretchers*, so arranged that each header is accurately aligned with a header two courses below. It was introduced at the highest fashionable level in the early seventeenth century, and gradually superseded *English bond* where appearance was important. |
| Flight hole | A hole in the roof or upper part of a dovecote through which pigeons could enter and leave. It can be of various shapes, but usually is just large enough to allow pigeons to pass through, too small for birds of prey. |
| Flight platform | A platform inside the roof of a dovecote on which the pigeons could alight before descending into the interior. |
| Gablet | A small gable, set back from one wall of a rectangular building. |
| Gauged bricks | Bricks accurately rubbed to shape, so as to fit a gauge. |
| Gault bricks | Facing bricks of fine-textured, pale grey fabric, very fashionable in the early nineteenth century. |
| Girt | A main horizontal timber connecting the principal posts at about mid-height. |
| Green bricks | Bricks during or after drying, but before firing. |
| Groundsill | The horizontal member on which a timber frame is built, usually mounted on a plinth of masonry. |
| Halved | Where two timbers cross, half the depth of each is cut out so that they are jointed together without exceeding the thickness of one timber. |
| Header | A brick laid with its length at right angles to the line of the wall, exposing the end. |
| Joggled | Said of two subsidiary timbers which are jointed into the same main timber from opposite sides, but not directly opposite each other. |
| Jowl | A swelling at the head of a post, strengthening the end joint to the timbers above. |
| Kiln-fired | Of bricks, fired in a permanent installation which allowed more accurate control of the process than the temporary *clamps* used near the site. |
| Loop | A narrow vertical slot in a wall for light, ventilation or defence. |
| Louver | A turret on the roof of a dovecote, usually with sets of parallel inclined boards about 6 inches apart, which allowed the pigeons to enter but which kept out the larger birds of prey. Also spelled 'lover' and 'lovre' in contemporary sources. The spelling 'louvre' is a modern corruption, influenced by the irrelevant Louvre in Paris. |
| Nest-box | An enclosed space provided for pigeons to nest in, constructed separately from the main structure. |

| | |
|---|---|
| Nest-hole | A cavity formed in the main structural fabric for pigeons to nest in. |
| Nogged | See *Brick nogging*. |
| Normal assembly | A term devised by Cecil Hewett in 1961 to describe the commonest form of tying assembly in English timber-framed buildings.[8] The tie-beam is laid over the wallplate and is secured to it by a lapped dovetail joint. If over a post it is also secured to a tenon rising from the *jowl*. |
| Ogee | A double-curved line or profile made up of a convex and a concave part. |
| Perching ledge | A ledge on the outside of a dovecote, provided for the pigeons to perch on out of the wind. Some are inclined so that the droppings are washed off by rain. |
| Pipe | A protective device inside a dovecote against birds of prey, described at Great Thurlow and elsewhere. |
| Potence | The modern term for a revolving frame which supported a ladder, allowing the pigeon-keeper to search the nesting places without descending. The word was introduced into English in 1887 by R. S. Ferguson, quoting from Viollet-le-Duc's description of French dovecotes. |
| Pugged | In the finer qualities of bricks the clay was passed through a horse-driven pugging mill in which rotating blades broke up the lumps and reduced the material to a smooth even consistency before moulding. It was in common use by the eighteenth century, earlier for bricks of the highest quality. |
| Soffit | The under-surface of any architectural structure or component. |
| Sprocket | A wooden wedge nailed to the foot of a rafter to cant the lowest course of tiles to a flatter pitch than the remainder. |
| Squab | A young pigeon, still unable to fly. |
| Stretcher | A brick laid with its long side parallel with the length of the wall. |
| Stretcher bond | A bond in which all bricks are laid longitudinally. |
| Stud | A subsidiary upright timber in a wall, between the main posts. |
| Wallplate | The horizontal timber along the top of a wall. |
| Wattle and daub | Wattle is a lightweight framework made of staves and pliant sticks to which the daub was applied like plaster. Daub is basically mud (sometimes with animal dung added), trodden with water and vegetable fibre to a plastic consistency. |
| Wattle fixings | Holes or grooves in a timber frame which secured the staves of wattle. Even when the wattle and daub has been removed the fixings remain, showing that the panel was formerly filled with wattle and daub. |
| Weathered, weathering | Bricks or freestone shaped to form an inclined surface, usually at 45 degrees, to shed rainwater. |

# Chapter 1
# How dovecotes were used

## WHAT WAS A DOVECOTE FOR?

The main purpose of a dovecote was to provide a luxurious supplement to the diet of the wealthy, who (despite what Cooke and others have written) already had plenty of fresh meat throughout the year. The art of the pigeon-keeper was to exploit the breeding behaviour of domesticated versions of *Columba livia*, wild rock doves, and to take the young birds for food. In nature these birds congregate on high cliffs, perching on rocky ledges and nesting in the dark recesses of caves. (They are quite different from wood-pigeons, which are tree-nesting birds). Every dovecote, however architecturally distinguished, is effectively an artificial version of that natural habitat.

## THE PRODUCTION OF MEAT

*Columba livia* differ from most other birds in breeding repeatedly through two-thirds of the year. Dovecote pigeons produced four to seven clutches of two eggs a year, and incubated them for seventeen days. Both adult birds fed the young, called *squabs*, for about thirty days, by which time they were ready to fly. In nature the parent birds would then drive the squabs off the nest. In domesticity the pigeon-keeper took the squabs shortly before they were ready to fledge, wrung their necks, and delivered them to the kitchen or the market. They weighed about one pound; as the flying muscles had never been used the meat was exceptionally tender, and could be cooked by roasting on a spit before an open fire.

## THE SEASONAL NATURE OF THE SUPPLY

It is often claimed that dovecotes provided a source of fresh meat in winter, but household accounts from all periods show that this food was eaten mainly from early April to early November. The largest numbers of squabs were drawn from the dovecote in August, September and October, when other meat was plentiful. For example, in 1412-3 Dame Alice de Bryene's large household at Acton, Suffolk, consumed 100 pigeons in April, 238 in May, 54 in June, 54 in July, 336 in August, 340 in September, 275 in October, and 147 in November - but none in December, January or February, and only 4 in March.[9]

Occasionally some pigeons bred in winter, but as a food supply the quantity was negligible. A very wealthy household such as that of the Earl of Salisbury in 1634 could buy small numbers of winter squabs, and so could enjoy this delicacy out of season. Their rarity was expressed by the price paid, four times as much in February as in October.[10] Where a household ate only the products of its own dovecote, as most did, pigeons effectively disappeared from the menu in mid-November, and became available again at Easter.[11]

It is impossible to appreciate the social significance of dovecotes unless their association with a luxurious way of life is understood.

## STOCKING THE DOVECOTE

In 1846 the naturalist Charles Waterton built a new dovecote near his house and demolished an old one some distance away in his park. He described how easily the new building was stocked. On a March day while the pigeons were out feeding he blocked the entrances to the old building and opened the new one; the pigeons took to it immediately, and by November of that year they had produced several hundred squabs.[12] Some dovecotes could be stocked from an earlier one on the same property, but others had to be stocked from another flock. How this was done was described by Leonard Mascall in 1581.[13] Young pigeons which had not yet flown were brought to the new dovecote, confined within it for two weeks, and fed on cummin and other specially favoured seeds. They were first released late in the day, so that they would not fly far before the light faded. Each day they were released a little earlier, until they had become fully attached to their new home.

## FEEDING

In traditional practice, for most of the year the pigeons were left to find their own food by ranging over the surrounding land. In the depth of winter they were fed with cereals and pulses, and also for a period between mid-summer and the pea harvest, when there was little food on the ground for them. This was called the 'benting time', because otherwise they would have had to subsist on bents, wild grasses. Pigeons will eat green vegetation and small snails, but primarily they are seed-eating birds. On meadow, pasture, and heath they could feed on natural seeds without doing any harm. In farmsteads and on cultivated land they could often find some spilled or wasted grain. They would feed on the seeds of wild plants, and if they were allowed to they might also feed on newly sown seeds and cultivated crops. To the modern reader this may seem wasteful (and certainly modern farmers regard pigeons as harmful), but we should remember that children were employed to drive all birds off the crops at seeding time and before harvest, when they were most vulnerable. Traditionally the pigeons were fed on the ground, as shown in medieval illustrations.[14] In the early nineteenth century J. C. Loudon advised that they should be fed inside the cote; special feeding platforms were built at that time (as will be described at Flixton Hall), in accordance with his advice.[15] In 1807 Thomas Rudge estimated that a bushel of corn a day was required to feed a flock of one thousand.[16]

## MAINTAINING THE BREEDING STOCK

The pigeon-keeper 'allowed one flight to fly' - that is, he allowed one month's batch of squabs to grow to maturity to become breeding stock. Females can lay within 120 days of being hatched, so some pigeon-keepers retained the first flight of the year, because they would produce young the same year; others 'let the harvest flight fly'. The decision on which batch to retain for breeding depended primarily on the local food supply, but like everything else in livestock management it was a personal choice based on experience and judgement. The objective was always the same, to maintain a young and healthy breeding stock which would produce a sufficient supply of squabs for the household or the market.

At the end of the breeding season surplus males and old birds which had ceased to breed frequently were culled. This too was described by Leonard Mascall in 1581.[17] At night when all the pigeons were inside, a net was spread over the *louver*. One man drove the pigeons out while another on a ladder caught them in the net, examined them, and either wrung their necks or released them. It is unlikely that the culled birds were thrown away, but their meat was tough, and could be made palatable only by prolonged stewing. Only the squabs were considered to be of value.

**MINERAL SUPPLEMENTS**
Like dairy cows, pigeons need a mineral supplement. By the eighteenth century pigeon-keepers had worked out that the main requirements were salt, calcium, and grit, but their less scientific predecessors prepared what was called a 'salt-cat' from ingredients which sound disgusting to the modern reader. The 'salt-cat' or mineral supplement was hung in the dovecote for the pigeons to peck from. If it was not provided they would peck lime mortar from buildings. Early descriptions indicate that a good deal of superstitious ritual went into the preparation of the salt-cat, and into charms intended to drive off predators.[18]

**USES OF THE MANURE**
Pigeon droppings formed a valuable by-product. The traditional practice was to allow the dung to accumulate on the floor of the dovecote, and to remove it in winter to avoid disturbing breeding birds. The main use was as manure. In 1577 Thomas Tusser (who farmed at Cattawade near the Stour estuary) set out the tasks for each month of the farming year, and for January wrote:

> *Feed dove (no more killing), old Dove house repaire,*
> *Save dove dong for hopyard, when house ye make faire.*[19]

Eighteenth-century and later writers favoured removing the dung more often. They claimed that one load of pigeon manure was worth ten loads of animal manure; therefore it was economic to cart it to the most distant fields. Also, it was sold to tanners for treating hides.

From 1560 pigeon dung was collected for making saltpetre. Until then England had been wholly dependent upon imported gunpowder. Queen Elizabeth's Privy Council paid a Friesian refugee, Gerhard Honrick, to reveal the Continental technology for making saltpetre. Agents of the Crown were given compulsory powers to collect pigeon manure from dovecotes, and to dig up the earth floors, to be used in the production of saltpetre for gunpowder. In 1625 a further order forbade owners of dovecotes from paving the floors; nothing but 'good and mellow earth' was to be used. The abuse of these powers by a private monopoly appointed by Charles I became one of the matters of contention between Parliament and Crown, and they were abolished in 1641.[20] From that time most saltpetre was imported.

## THE USE OF LIME-WASH

In all countries and at all historic periods the upper parts of dovecotes have been whitened with lime-wash or white plaster. Ovid wrote (in Waterton's translation):

> *See, to the whitewash'd cot what birds have flown*
> *While that unwhitewash'd, not a bird will own.*[21]

Columella wrote: 'The whole place and the pigeon-cells themselves ought to be finished off with white plaster, since birds of this kind take a special pleasure in that colour'.[22] No one knows why pigeons are attracted to whitened places, but one might speculate that in nature the accumulations of white guano round well-established nesting sites led rock doves to associate a whitened area with security, a place which had been occupied successfully for many generations. The practical pigeon-keeper knew from the accumulated experience of his predecessors that it was good practice to use lime-wash on the dovecote, particularly round the pigeon entrances. (In the Appendix Dr. Hoppitt records a dovecote at Sapiston being whitened with lime in 1337).

Nineteenth-century writers advocated lime-washing the interiors of the nesting places to control infestation by insect parasites.[23] Today we often find the remains of lime-wash inside dovecotes, and occasionally traces on the outside, but mostly it has been washed away by weather.

## THE SCOPE OF THIS STUDY

This study is concerned with historic dovecotes and with traditional pigeon-keeping, where the main purpose was to produce palatable meat. Dovecotes are understood to be special-purpose buildings designed and used within that tradition. Pigeon lofts inserted into existing buildings, or lofts erected mainly for ornamental or racing pigeons, are outside the scope of this study.

# Chapter 2
# Who could own a dovecote?

Pigeon-keeping was already ancient by the Roman period; it was described by Varro in the first century B.C., and in more detail by Columella in the first century A.D.[24] It is not known whether the Romans kept pigeons in Britain, but effectively the practice was introduced by wealthy Norman lords - the same people who introduced fallow deer and rabbits from abroad, and formed fenced parks and warrens to supply their own tables. In the early Middle Ages owning a dovecote was a prerogative of powerful lordships, comparable with the ownership of mills; only the wealthy could build them. Corporate landlords such as monasteries and colleges exercised the same manorial rights, and some parish priests obtained the same privilege. There was an informal relationship between the size of the estate and the number of birds kept - the larger the estate, the greater the number of nesting places provided.

The number of dovecotes increased continuously as more landowners took up the rights of lordship. Following the dissolution of the monasteries and chantries between 1536 and 1548 about one-fifth of the land of England passed into secular hands; many of the new landowners built dovecotes on these estates, irrespective of whether that right had been exercised there earlier. This became one of the complaints made in Kett's rebellion in 1549, a rising mainly in Norfolk, in which Suffolk men were involved.[25] Despite some criticism that the large number of dovecotes already constituted a nuisance they continued to increase, and by 1655 one observer estimated that on average there were three dovecotes in each parish. That is the basis of the much-quoted (but very dubious) statistic that by then there were 26,000 dovecotes in England.[26]

C. Northcote Parkinson observed that organisations erect their most impressive buildings when their power is already in decline.[27] His argument might well be applied to dovecotes. In the early Middle Ages feudal magnates exercised real power, but their dovecotes were plain, functional buildings - although of massive construction. As the powers and privileges of landlords declined dovecotes became more important as symbols of high status (or of asserted status), rather like coats of arms. Increasingly dovecotes were treated with architectural refinement, and were sited where they were most conspicuous. They were built of the best materials, in the most fashionable style of the day; and in the eighteenth century some were designed as 'eye-catchers' in planned landscapes. Effectively they encapsulate the history of English architecture in microcosm.

The right to build and use a dovecote was limited only by common law.[28] Until the early seventeenth century it was a privilege reserved to lords of manors and parish priests. The common law was modified by an important case in 1619; from that time any freeholder was entitled to build a dovecote on his own land.[29] By the nineteenth century many tenant farmers had become pigeon-keepers. At that time the pigeon-loft was often combined with a lower storey devoted to another agricultural purpose, or a pigeon-loft was inserted in the roof of an existing building.

# Chapter 3
# How dovecotes were designed

Unlike other domesticated creatures dovecote pigeons were free to leave at any time. Sometimes they left *en masse* to join a flock from another dovecote, or even to join a wild flock, for reasons which were not always apparent to the owner.[30] Consequently they acquired a reputation for being fickle birds, which had to be wooed by providing favourable conditions. The difference between one dovecote and another in the same region often reflects the owners' beliefs about what was most attractive to the pigeons.

## PIGEONS' INSTINCTIVE RESPONSES TO BIRDS OF PREY
In level flight pigeons fly so fast that they can usually escape from birds of prey (except peregrine falcons, which fly even faster), but they are vulnerable to attack when perching, or when taking off and gaining height. Much of their behaviour is an instinctive response to this hazard. They choose high places on which to perch so that they can see birds of prey approaching, and they stay together in a flock because the more eyes there are, the sooner a predator will be seen. Also, by diving from a high point they can pick up speed rapidly, and get away. When alarmed by a bird of prey their natural instinct is to fly fast, weaving and swerving when the pursuer approaches. They 'will dive headlong into any suitably sized hole or crevice with a complete lack of any of the caution otherwise shown when entering an unknown cavity'.[31]

## THE PERCHING BEHAVIOUR OF PIGEONS
Like the wild rock doves from which they are descended they are well adapted to perching on steep surfaces and narrow rocky ledges, so they have no difficulty in perching on roofs. When not out feeding, and particularly in early morning and evening, they choose to perch in the sun. In strong winds they find a sheltered place to perch - on the lee side of the roof, or behind a dormer or pinnacle, or on a string course or ledge on the lee side of the building.

## DESIGNING THE EXTERIOR
It is unlikely that those who owned or built historic dovecotes analysed bird behaviour in the same way as modern naturalists, but by observation and experience they learned what they most needed to know. They observed that pigeons naturally choose to perch on the roofs of tall buildings, so they built dovecotes high. They saw that in windy conditions pigeons try to get out of the wind, so they provided sheltered perching places. By looking at other dovecotes and exchanging information with other pigeon-keepers they came to recognize that some roof shapes are better for the pigeons than others. The ideal roof for a dovecote has inclined surfaces facing in all directions, so that at all times of day part of it is facing the sun, and in all wind conditions part of it is in shelter. Builders who worked on roofs did not need to

know the word 'micro-climate' to understand that some roofs provide a warm still area even in the strongest wind. At highly exposed sites, particularly near the coast, they provided *perching ledges* high on the walls.

## DESIGNING THE INTERIOR
When the owner or master builder took design decisions about the interior he did not set out to reproduce a natural cave, but experience told him that pigeons choose to nest in dark places. Some early dovecotes were made without windows, deriving sufficient light from the opening at the top through which the birds entered. If windows were provided they were small. A pigeon-keeper would notice that some nesting places were always left unoccupied, and would develop ideas about why some were more attractive than others. In England many dovecotes have *nest-holes* or *nest-boxes* which are L-shaped in plan; the pigeon entered a small aperture and occupied a chamber to one side. This shape has the merit of being darker inside than a plain recess.

Traditionally the builder provided as many nesting places as the structure would accommodate - although in the eighteenth and nineteenth centuries other factors began to operate, as will be reported. Masons who had learned how to construct nest-holes in stone buildings had to start learning all over again when bricks came into common use, for unlike stone, bricks came in standard sizes. They had to decide how many courses of bricks to allow for each tier of nesting places, and it is clear that different builders came to different conclusions. In 1682 the Rector of Clayworth in Nottinghamshire described how he decided to increase the height of the nest-holes after the first few tiers had been built.[32] In 1698 Roger North discussed in some detail how he designed the roof of his new dovecote at Rougham, Norfolk.[33] Otherwise, few pigeon-keepers committed their thoughts to paper. Some writers offered general advice about pigeon-keeping (and they are quoted here), but in many cases we can only speculate about how a particular design decision was taken. The buildings themselves, where they survive in sufficiently unaltered condition, provide the best evidence of what factors were considered when they were designed.

## PROTECTION AGAINST BIRDS OF PREY
A bird of prey which penetrated to the interior of a dovecote would terrify the pigeons (Shakespeare used the phrase 'like an eagle in a dovecote' to express this effect).[34] Various devices were developed to keep birds of prey out. Traditionally the pigeon entrance took the form of a louver, a turret on the roof fitted at the sides with parallel boards inclined at 45 degrees, about 6 inches apart. The pigeons could alight on the boards and pass through easily, but the larger birds of prey could not. (A well-restored louver may be seen at the National Trust dovecote at Ightham Mote, Kent, and some others survive unrestored, although usually in fragmentary condition). In the eighteenth century many architects embellished dovecotes with glazed lanterns, but they provided horizontal slots 6 inches deep at the base for the pigeons to enter. Other dovecotes had canopies over the pigeon entrance, which prevented hovering birds of prey from attacking from above.

Sparrowhawks and owls were small enough to slip through the slots of the louver or lantern, and could fly under the canopy, so another device was developed, called a *pipe*, which will be described at Great Thurlow, Polstead and elsewhere. Whether owls were really harmful in dovecotes is a matter of controversy. Contemporary reports did not distinguish between the different species of owls; one resolution of the conflicting accounts is that barn owls did not predate upon squabs, but that tawny owls did.[35]

If a sparrowhawk or other bird of prey managed to get inside the building the pigeon-keeper tried to ensure that there was no place where it could perch or take refuge. Pigeons can perch easily on the *alighting ledges* provided in front of the nesting places, but they were deliberately made too narrow for tree-nesting birds. As far as possible the roof was constructed without free-standing tie-beams on which a predator could settle. As first built some of these roofs were insufficiently tied against expanding forces, and over the years they have spread. Tie-beams and other timbers have been introduced to prevent further movement, but examination shows that they were not present in the original construction.

The windows of dovecotes were protected against birds of prey by iron grills or closely-spaced wooden slats. These tend to decay, but some surviving evidence will be reported here.

## PROTECTION AGAINST OTHER PREDATORS

In warmer climates pigeon-keepers had to protect dovecotes against reptiles, but this was not a problem in Britain. The early literature is confused on the point, because English gentlemen educated mainly in the classics respectfully repeated advice on pigeon-keeping by the Roman author Varro, who had written for Mediterranean conditions.[36] In Britain, until the eighteenth century, the only quadrupeds which predated upon dovecotes were polecats (and the related species, martens, stoats and weasels, which were not clearly distinguished in contemporary literature) and cats (domestic and feral). In 1347 at Standon in Hertfordshire the annual yield of a dovecote fell from seven shillings to two shillings, and the auditors accepted the explanation that 'it was destroyed for most of the year by polecats which were caught afterwards and the doves were re-stocked'.[37] A water barrier for defence against these predators will be described at Aspall Hall. In the modern literature about dovecotes there are many references to what are alleged to be protective devices against 'climbing predators', but on further inspection they usually prove to be perching ledges, or string courses and cornices similar to those which embellished houses and public buildings of the same period.[38]

## THE INTRODUCTION OF BROWN RATS

Rats had been present in Britain since the Roman occupation, but these were black rats, *Rattus rattus* - a different species from the destructive rats with which we are more familiar, and entirely different in behaviour. Black rats ate fruit and grain. They did not predate upon livestock, and they could not gnaw through building materials. They were pests in the towns, but they were never widespread in the countryside. Early farming literature shows that they were not regarded as a significant nuisance on farms.[39]

That situation changed entirely with the introduction of another species, brown rats, *Rattus norvegicus*. They were indigenous in eastern Asia, but in the eighteenth century they spread overland to Russia, and were carried to Britain by shipping from Baltic ports. The evidence has been discussed in detail elsewhere, but in brief, it indicates that this species entered Britain at the port of London in the decade 1720 - 1730.[40] In 1735 one English pigeon-keeper advocated elaborate defensive measures against predatory rats, but at that date he could not distinguish between the different behaviour of the two species.[41] In 1740 another wrote of rats as essentially an urban problem:

> *The City's odious to the harmless Dove,*
> *Business suits ill with Innocence and Love.*
> *Beside, in Towns the Rat's insidious Kind*
> *Too often in the Cote an Entrance find;*
> *Break the thin Eggs, and make with fruitless Pain*
> *Th'eluded Mother sit whole Months in vain.*[42]

From London the new species spread up the river system, and was carried by shipping to other ports. By 1748 they had penetrated far up a tributary of the Thames to the Chilterns, and by 1754 they were reported at Wansford on the River Nene.[43] It is likely that they colonized the Suffolk rivers at much the same time. By streams and drainage channels they would have reached most Suffolk farmsteads by the 1750s, and even the remotest by the 1760s.

To pigeon-keepers their arrival was often a disaster. Timber-framed buildings were highly vulnerable, for brown rats could burrow under shallow foundations, or could gnaw through panels of *wattle and daub*, or a cladding of lath and plaster. They could gnaw through doors, unless protected by iron plates, or by a dog kennelled nearby. Where one rat found its way in it left traces which others followed. Within a short time the building would be thoroughly infested, and the whole breeding stock would be destroyed. Probably many timber-framed dovecotes were abandoned as useless from that time, but others were adapted to the new hazard, as will be described in the survey.

Some brick and stone dovecotes built earlier were inherently resistant to brown rats, because the foundations were too deep to be undermined, and the fabric too hard to be penetrated. Others were altered - by raising vulnerable doorways above the reach of rats, by blocking the lower tiers of nest-holes, or by removing the lower nest-boxes. New dovecotes built after the mid-eighteenth century were designed from the outset to be rat-proof.

# Chapter 4
# The decline of dovecotes

From the mid-seventeenth century some writers condemned pigeon-keeping as causing a great waste of cereals, and as a nuisance to other farmers, but dovecotes continued in use, and new ones continued to be built. This continued until the last few years of the eighteenth century, but in the French revolutionary wars major economic changes began to affect pigeon-keepers. The price of wheat rose to unprecedented levels, and progressive agriculturists calculated that the corn which pigeons consumed was worth more than they produced as meat and manure. Their calculations were crude and unscientific, but they were convincing in their time. From 1794 many of the county reports to the Board of Agriculture described pigeons as wasteful, and in Norfolk and Kent they added that dovecotes were being taken down.[44] By 1825 J. C. Loudon could state that pigeons were 'scarcely admissible in professional agriculture, except in grazing districts, where the birds have not so direct an opportunity of injuring corn . . . Pigeons are now much less cultivated than formerly, being found injurious to corn fields, and especially to fields of peas. They are, however, very ornamental; a few may be kept by most farmers, and fed with the common poultry'.[45]

Pigeon-keeping on the traditional pattern declined sharply, but on some farms it was replaced by a small-scale operation in which the pigeons were fed in the yard to discourage them from ranging widely over neighbours' fields. Where suitable, some dovecotes were altered by inserting a floor at mid-height, retaining the upper part as a pigeon-loft while converting the lower storey to a stable, gig-house or other purpose. At others the nest-boxes were rebuilt at that time to provide better conditions for a smaller number of pigeons. It became important not to have surplus nest-boxes, to avoid attracting more pigeons than the owner could feed on his own land.[46]

**TRAP-SHOOTING**

The modern sport of clay pigeon shooting is a humane substitute for the earlier practice of releasing live pigeons as targets. This was a new development of the early nineteenth century; shooting clubs were formed, and they expanded rapidly, generating a large demand for live pigeons. Unscrupulous suppliers fulfilled the demand by stealing adult birds from working dovecotes. Charles Waterton in Yorkshire described their methods in 1839, and said that the dovecotes in his neighbourhood had been robbed repeatedly, often of their entire stock.[47] Isolated dovecotes became impossible to protect from thieves, so most new ones were built within enclosed yards, or as lofts over occupied buildings. Some pigeon-keepers were able to profit from the new 'sport' by breeding pigeons for sale to the shooting clubs.[48] Pigeons which the guns missed simply flew home, and were sold again!

## THE END OF THE WORKING TRADITION

By the middle of the nineteenth century most dovecotes were passing out of use; many were already derelict. In 1887 Ferguson reported that within living memory coaching inns along the Great North Road had kept pigeons to provide a ready meal for travellers, but their trade was destroyed by the growth of railways - a development of the 1840s. By the 1880s the practice of keeping pigeons for meat had ceased to be part of the common culture of the countryside, and he was unable to explain some features of the dovecotes he described. Watkins assembled a list of 36 dovecotes in Herefordshire which had been pulled down by 1890.[49] Some curious propositions were put into circulation by writers who had not studied the earlier farming literature, and continue to be repeated.[50] The explanations given here are derived from accounts written while pigeon-keeping for meat was still widely practised and understood.

# The Survey of Suffolk Dovecotes

It is convenient to arrange the descriptions of Suffolk dovecotes in groups according to the main structural material - timber framing, brick, and finally stone. Those in the first group are not presented in strict chronological order, for it may be more helpful to the reader to discuss some features which have survived in good order before looking at other buildings where they are less complete. In the second and third groups the buildings are arranged in chronological order, but in some cases the dating of the buildings is only approximate. Chapter 8 describes 'Three Oddities', and Chapter 9 analyses the results of the survey, and draws some comparisons and conclusions.

### DIMENSIONS, DISTANCES AND ORIENTATIONS
Dimensions are given in the traditional English units in which these buildings were built. For consistency distances are expressed in yards and miles:
- 1 inch = 25.4 mm
- 1 foot = 12 inches = 0.305 metre
- 1 mile = 1.61 km

The orientation of each building is stated exactly in the introductory paragraph, and in the caption of each exterior photograph. Subsequently it is sufficient to describe each feature in terms of the nearest cardinal point.

### STATUTORY LISTING
Unless otherwise stated all the buildings described here are Listed as Buildings of Special Architectural or Historic Interest, Grade II - the category which includes some 92 per cent of all Listed Buildings.

### A NOTE ABOUT GROUND LEVELS
Ground levels tend to rise over a long period owing to the formation of humus and the deposition of waste material. In ancient buildings the original floor level is often well below the surrounding ground. In this survey it has not always been possible to identify the original ground level, so heights are given from present ground level.

# THE DOVECOTES OF SUFFOLK,
## with district boundaries and main towns

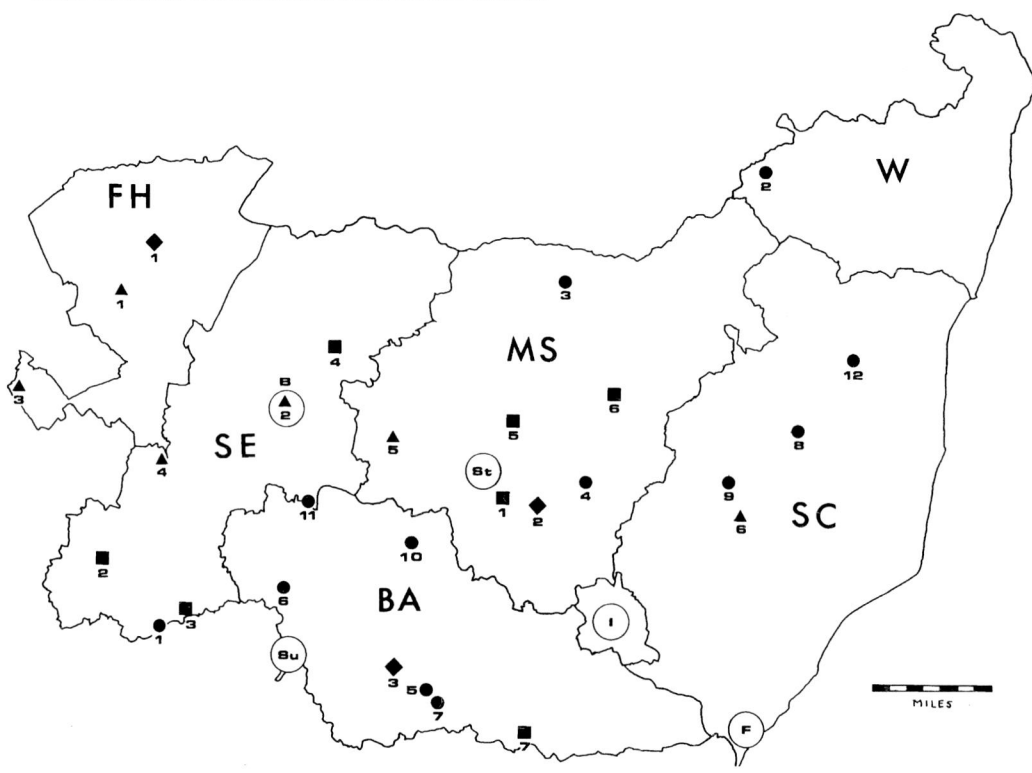

**KEY:**

■ TIMBER-FRAMED DOVECOTES:
1. Great Thurlow Hall, Great Thurlow
2. Badley Hall, Badley
3. Church Farm, Clare
4. Redcastle Farm, Pakenham
5. Chapel Farm, Gipping
6. Aspall Hall, Aspall
7. Claycotts, East Bergholt

● BRICK DOVECOTES:
1. Stoke College, Stoke-by-Clare
2. Flixton Hall, Flixton by Bungay
3. Goswold Hall, Thrandeston
4. Crowfield Hall, Crowfield
5. Polstead Ponds Farm, Polstead
6. Kentwell Hall, Long Melford
7. Scotland Place, Stoke-by-Nayland
8. The Park, Great Glemham
9. Four Bridges, Letheringham
10. Hitcham House, Hitcham
11. Coldham Hall, Stanningfield
12. Cockfield Hall, Yoxford

▲ STONE DOVECOTES:
1. Church Walk, Mildenhall
2. Abbey Gardens, Bury St. Edmunds
3. Marsh Stables, Exning
4. Ousden Hall, Ousden
5. Home Farm, Drinkstone
6. Home Farm, Pettistree

◆ 'ODDITIES':
1. Eriswell Hall, Eriswell
2. Alder Carr Farm, Creeting St. Mary
3. Parsonage Farm, Boxford

**DISTRICTS AND MAIN TOWNS:**
BA  Babergh
B   Bury St. Edmunds
F   Felixstowe
FH  Forest Heath
I   Ipswich
MS  Mid Suffolk
SC  Suffolk Coastal
SE  St. Edmundsbury
St  Stowmarket
Su  Sudbury
W   Waveney

# Chapter 5
# Timber-framed dovecotes

**GREAT THURLOW HALL, GREAT THURLOW**  (TL 683 503)
This dovecote is in the upper Stour valley, 150 yards east of the Hall and the parish church, and slightly uphill from them, separated from the present farm buildings by a small paddock (Figure 1). There is a pond 5 yards to the north. The building is 24 feet square, 14 feet high to the eaves, with an off-centre door facing south-south-east, and an unglazed window obliquely opposite; other windows to north and south are modern. The plain-tiled roof is hipped with *gablets* to east and west, now boarded over.

### The brick plinth:
The *groundsills* are raised 1½ feet above present ground level on a brick plinth. The bricks are 9 x 4½ x 2½ inches, laid in lime mortar in *Flemish bond*, four courses rising 11 inches. The present floor is of concrete.

Figure 1: The timber-framed dovecote of Great Thurlow Hall, Great Thurlow, from the south.

### The timber frame:

The frame is of oak of high quality, and is almost intact, clad with cement render on modern lathing. There are no wattle fixings, so apparently it was clad with lath and plaster originally. The four *wallplates* are arranged at the same level, with long angle ties across the corners, tenoned and pegged to them. The north and south walls are built in three *bays*, the east and west walls in two bays. The corner posts have identical *jowls* which terminate in an elegant *ogee* curve; the intermediate posts are unjowled.

### The doorway and windows:

The doorway is against a main post and below a *girt*; it has been enlarged for later use, but jointing evidence shows that originally it was only 4 feet 10 inches high by 2 feet 1 inch wide, with a window of similar width above. The north window survives in original condition; it is 1 foot 6 inches high by 2 feet 3 inches wide, built against a main post and immediately above a girt, protected against birds of prey by a grill formed of horizontal iron bars ½-inch wide about 1 inch apart.

### The roof:

Principal rafters above the intermediate posts in the north and south walls form the gablets. *Collars* are *halved* across them to form their bases. Principal rafters above the middle posts of the east and west walls meet the collars. *Joggled* purlins are tenoned and pegged to the principal rafters, and common rafters of 4 x 2¾ inch vertical section are tenoned and pegged to them. A complete set of finely shaped *sprockets* survives.

### The nest-boxes:

The walls were lined from the outset with ¾-inch pine boards, some of which remain *in situ*. The nest-boxes have been removed, but fragments which remain, and stains on the boarding, show that they too were made of pine; there were 13 tiers, with 78 boxes to a complete tier. Allowing for those omitted at the door and windows there would have been nearly one thousand. They were 12 inches wide and 9½ inches from front to back; because the tiers were fitted between structural timbers the heights varied from 9 to 10½ inches. There was an alighting ledge of unknown width to each tier. Some 1¼-inch pine boards which formed the fronts of the nest-boxes have been re-used to build a gallery; the entrance holes were of inverted-U shape, 5 inches high by 4½ inches wide. Only the fronts of the nest-boxes were lime-washed.

### The flight platform and pipe:

Between the gablets is an original *flight platform*, from which a funnel-shaped *pipe* descends into the interior (Figure 2). This was an ingenious protective device against sparrowhawks, which could sometimes penetrate to the inside of a dovecote. Pigeons can fly vertically up and down, so they could pass through the pipe without difficulty, but sparrowhawks could not. A sparrowhawk which managed to drop down through the pipe into the building would be trapped there until it was found

Figure 2: Looking up at the flight deck and protective pipe at Great Thurlow.

and destroyed by the pigeon-keeper. The pipe is formed of one-inch pine boards, with the framing outside so as to leave the inner surfaces smooth. The upper aperture is 2 feet 3 inches square, opening out to 2 feet 7 inches square at the bottom. The depth is 2 feet 6 inches. Similar pipes will be reported in other dovecotes. They vary in their horizontal dimensions, but the depth is remarkably constant, indicating that pigeon-keepers had determined what depth provided most effective protection against sparrowhawks without impeding the pigeons.

*Dating:*
 This structure is typical of the late seventeenth century. The use of so much pine implies lavish expenditure, for at that period it was the best timber obtainable. It was imported from Baltic ports, and the carriage so far inland greatly increased its cost. The large number of nest-boxes indicates that this dovecote was built for a major estate.

**BADLEY HALL, BADLEY** (TM 061 558)
This is an isolated manorial site 1½ miles north-west of Needham Market, near

Figure 3:
The timber-framed
dovecote of Badley Hall,
Badley, from the west.

the redundant parish church of St. Mary, approached by long access roads from east and west. The dovecote is 50 yards east of the remaining part of the sixteenth-century manor house, and slightly downhill from it. There is an impressive early sixteenth-century barn 20 yards to the south, and a pond immediately to the east. The dovecote is 18½ feet square, 11 feet high to the eaves (Figure 3). The central door faces west-south-west (visible from the house). The plain-tiled roof is hipped, with large gablets to east and west, now boarded over.

### The brick plinth:

The bricks are 9¾ x 4⅝ x 2 inches, typically sixteenth-century in material and texture, laid in lime mortar in *English bond*. It is likely that the ground level has risen substantially in four centuries, so probably the groundsills were nearly two feet above ground originally - which has preserved them in unusually good condition.

### The timber frame:

The wall framing is ingenious, and probably unique (Figure 4). The sides of the nest-boxes are formed by great oak boards 14 by 2 inches which also serve as the only studding. Curved down-braces ¾-inch thick are tenoned and pegged to

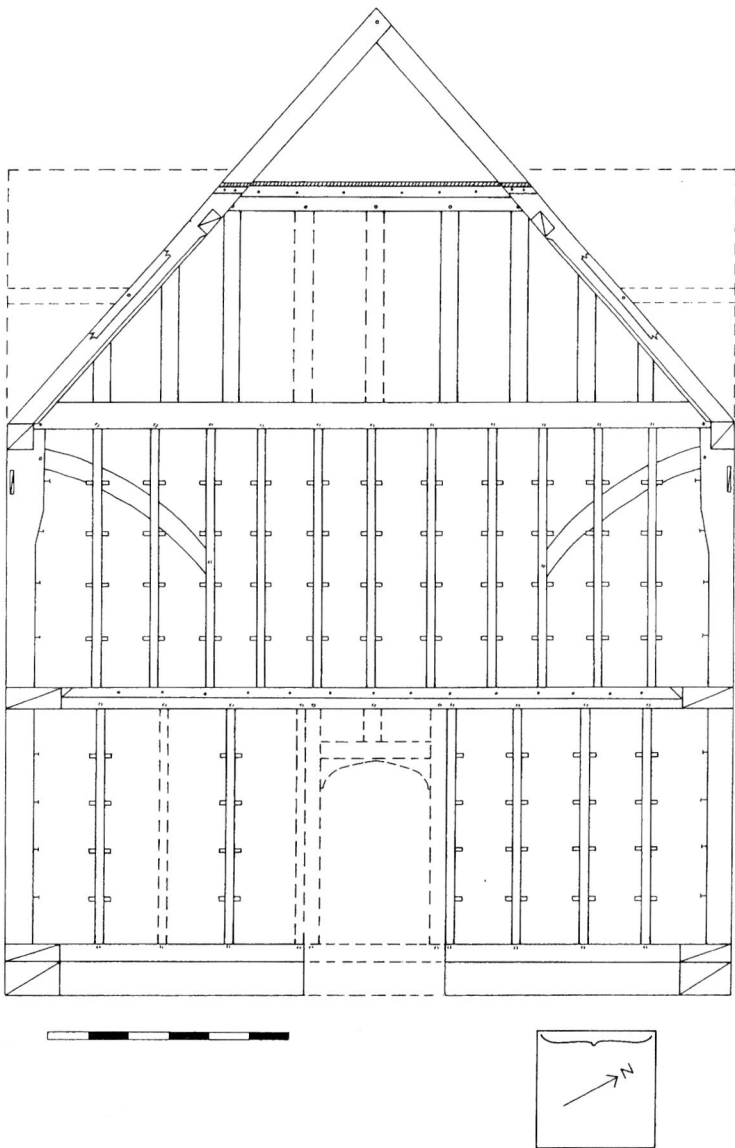

Figure 4: Vertical section of the dovecote of Badley Hall showing the west wall from the inside. Pecked lines indicate timbers which have been removed - to enlarge the doorway, to insert a window, and the dormers which were added after the original construction, and subsequently removed. (Scale in feet). By courtesy of Leigh Alston.

the jowled corner posts, each trenched across two of these 'studs' and terminating at the third. The posts and 'studs' are mounted on massive groundsills 14 x 5 inches supported one foot above present ground level on an original brick plinth. The cladding is of cement render on modern lathing. There are no fixings for wattle and daub, so probably the original cladding was of lath and plaster, or possibly of boards.

### *The doorway and window:*

The doorway is 7 feet high by 3½ feet wide; evidently it has been enlarged for use as a stable. There is a modern window beside it, but originally there were no other apertures in the walls.

### *The nest-boxes:*

The nest-boxes were formed by driving pairs of square pegs through ¾-inch auger holes in the vertical boards, projecting on both sides, and laying square oak baseboards on them. At the corners of the building these baseboards were neatly mitred and butt-joined (Figure 5). Most of them are now missing. This construction produced nest-boxes 13½ inches cubed, or in some places a little larger. Only one section survives of the vertical boarding which formed the fronts of the boxes, and even this is displaced from its former position. Originally these boards were nailed to the baseboards, with a pentagonal entrance hole 5½ inches high by 4 inches wide near the left side of each box. Wedge-shaped members 3 inches wide were nailed across the front to form an alighting ledge to each tier.

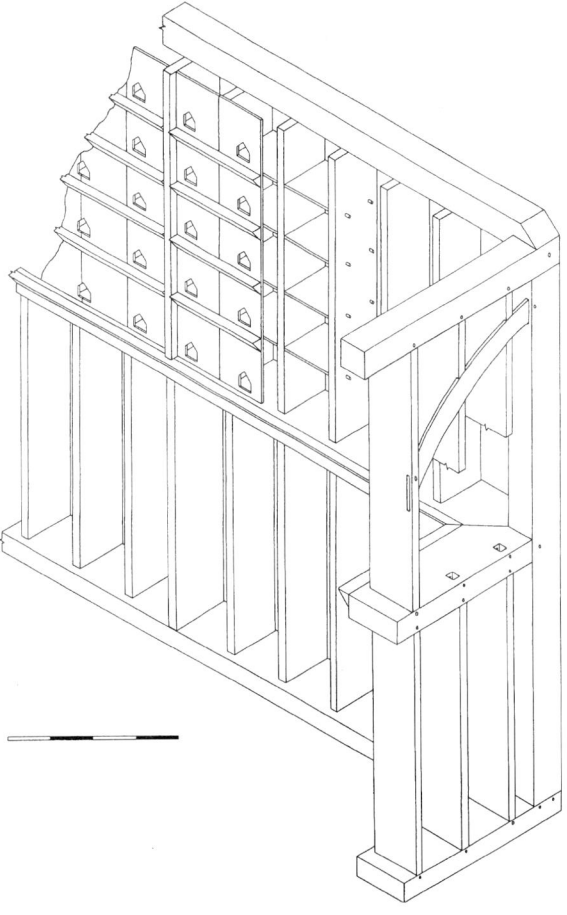

Figure 5: Isometric drawing of part of the dovecote of Badley Hall, showing how the nest-holes were formed with base-boards supported on pegs passed through the massive 'studding'. (Scale in feet). By courtesy of Leigh Alston

Great girts at mid-height project 2 inches beyond the fronts of the nest-boxes, and would have formed adequate alighting ledges of themselves, but similar triangular ledges are nailed to them too. There were 10 tiers of nest-boxes, with 48 in each complete tier, so allowing for those omitted at the door there would have been about 470 in all. This seems a small number for the high quality of the building and the importance of the site, but here the nest-boxes were unusually capacious.

## The roof and flight platform:

Full-length rafters rise from the north and south walls to form the gablets. Purlins are tenoned into them on all four sides. Shorter rafters rise from the wallplates and overlay the purlins, and are pegged in position. Another set of short rafters rises from the purlins, pegged to them to form the upper part of the roof, between the gablets. A complete set of sprockets remains *in situ*, itself a rare survival.

There is a flight platform of oak boards, but it has been rebuilt or strengthened at some time, for the oak joists supporting it are of 7 x 3 inch vertical section, which is characteristic of much later carpentry. The hole through which the pigeons descended into the interior is 1 foot 10 inches square.

## Defensive measures against brown rats:

The back of each nest-box has been lined with thin paving bricks, evidently for protection against brown rats. Many of the bricks have fallen away when the wooden bases were removed, but stains on the sides indicate their former presence. Inserting bricks into the nest-boxes would not have been possible in many dovecotes, but here they were so large that even when reduced there was still ample space for the pigeons.

This brick lining should not be confused with *brick nogging*, for here the bricks were inserted between the wooden baseboards at approximately 14-inch intervals. In the sixteenth century brick nogging was a fashionable feature of prestigious timber-framed buildings.[51] The bricks were arranged in geometrical patterns to be appreciated from outside, and they were held in position by shallow U-shaped or V-shaped grooves or depressions in the adjacent timbers. Here the absence of grooves confirms that the brick filling is secondary.

## Later developments:

Leigh Alston has found evidence that dormer windows have been inserted in the north and south pitches of the roof (as shown by pecked lines in Figure 4), and later removed. Originally the only light inside was that which filtered down from the roof. Dormers were rare in England before the early seventeenth century. In the long term they always lead to leakage and deterioration of the timber structure. These would have been removed at some later period when the roof was re-tiled, to correct the leakage and to simplify the tiling.

The building is now disused, but it has been used as a stable. In the lower half rough poles have been nailed from the girts to the groundsills to form an inner lining, and a wooden trough is still present against the west wall. It is not clear whether a floor has ever been inserted to form a pigeon-loft. In most buildings this would leave clear evidence, but here joists could have been laid across the projecting girts, and subsequently removed, without leaving any trace.

## Dating:

Dating such an unusual building presents problems, for there is really nothing with which to compare it, but an origin in the later sixteenth century seems most likely. The lavish use of large oak timbers may be more an expression of high manorial status, and the possession of sufficient woodlands, than an indication of a medieval origin.

### *Additional information:*

The manor house was built in the 1520s or 1530s for Edmund Poley (c. 1486 - 1549), and remained in the Poley family until 1735; there are many Poley memorials in the church. In 1674 tax was paid on thirty hearths. A map of 1741 shows that the house was formerly of quadrangle plan, of which the present building is only the south-western quarter.[52]

There is a large ash tree 7 yards north-west of the dovecote. Measurement of the girth shows that it is only about seventy years old, so it has been allowed to grow there since the dovecote passed out of use.

## 'SOUTH SUFFOLK A'

In two cases the owners of dovecotes allowed the author to examine and record their dovecotes, but asked that they should not be identified in a publication, as they wished not to be bothered by visitors. Readers who recognize the descriptions are asked to respect their wishes. The dovecote which will be identified as 'South Suffolk A' is isolated in a paddock 60 yards north-east of the house, and slightly downhill from it (Figure 6). There is a pond 20 yards to the north. The building is 17½ feet square, 12 feet high to the eaves, with a door at one end of the side facing

Figure 6: The timber-framed dovecote identified as 'South Suffolk A' from the north-west.

south-south-west (visible from the house). There are high windows in that side and the east side (the latter blocked externally). The plain-tiled roof has large gablets to east and west, now boarded over.

### *The timber frame:*

The oak frame has remained almost unaltered, and most of the original wattle and daub filling is present too. The corner posts are jowled for *normal assembly*. The oak studs are about 6 inches wide, 10 inches apart. No bracing is visible from inside, so the braces must be trenched outside the studs, concealed by the daub. The external cladding is of modern render. Some alterations have been made necessary by rising ground levels: the brick plinth has been rebuilt with eighteenth or nineteenth century bricks, and re-used oak timbers have been substituted for the original groundsills.

### *The doorway and windows:*

The original doorway was only 4 feet high by 2 feet 1 inch wide, rebated for a door opening outwards. To compensate for the rising ground level the sill has been raised by 10 inches, and the door-head has been raised by the same amount, so the doorway is still the same size as it was when built. The windows are 1 foot 8 inches high by 2 feet 2 inches wide; they are not original. So much of the timber structure is concealed that it is not possible to be certain, but it seems that there were no windows originally. The present windows each have a central diamond mullion and holes for a missing vertical bar to each side. If they had been formed as unglazed windows one would expect two mullions in this width, spaced about $6\frac{1}{2}$ inches apart. They are more typical of early glazed windows, in which a panel of leaded glass about 11 inches wide was fitted to each side of the central mullion, additionally supported on the inside by vertical bars (now missing, but leaving the holes where they were fitted).

### *The nest-boxes:*

The nest-boxes are made of wattle and daub (the wattle of thin coppice sticks, the daub of chalky clay), supported on cleft oak pegs driven into $1\frac{1}{2}$-inch auger holes in the posts and studs (Figure 7). They vary in size, but on average they are 14 inches cubed, each with an entrance hole 6 inches square set off to the right, and an alighting ledge 5 inches wide to each tier. There are now five tiers, with 43 to a complete tier; two lower tiers have been removed, leaving those above supported on inclined sticks. A few near the door have been removed, and some in the top tier are damaged. Allowing for those omitted at the door and windows there were 283 nest-boxes when this assembly was complete, of which about 180 remain intact. Some have been repaired with hand-made bricks and clay daub of a darker colour. No lime-wash is apparent.

However, these are not the original nest-boxes. Pairs of auger holes of $\frac{7}{8}$-inch diameter in the studs and posts at 9-inch vertical intervals indicate that a different pattern of nest-boxes was present earlier. No evidence of their form or construction remains, but on the assumption that they were made of one-inch boards and that

Figure 7: The south wall of the dovecote identified as 'South Suffolk A', with nest-boxes of wattle and daub. Note the two sets of auger holes in the studs - single holes of large diameter, and pairs of holes of smaller diameter. Only the lower part of the wall is brick-nogged; elsewhere the original filling of wattle and daub remains *in situ*.

each box was 12 inches wide there would have been 54 in each complete tier. There is sufficient height for 14 tiers. Allowing for those omitted at the door there would have been well over 700 nest-boxes.

### *The roof, flight platform and pipe:*

All the original oak rafters, of horizontal section, are present. Collars are tenoned across two pairs to form the gablets. The flight platform is original, and the greater part of a pipe remains *in situ* (Plate 1a, opposite page 42). Each side was made of two tapering boards to form a square funnel diverging at the bottom, similar to that at Great Thurlow; two of the boards have dropped away.

### *Defensive measures against brown rats:*

The spaces between the studs have been filled with brickwork to a height of 4 feet. The lowest tier of nest-boxes is 4½ feet above the earth floor. This would be sufficient to raise them above the reach of rats, provided that there was nothing on which they could climb. The present inclined struts are later insertions.

*Dating:*
The timber structure (or what is visible of it) is characteristic of the sixteenth century. The windows were inserted later, towards the end of the century or early in the seventeenth century. The doorway was raised in the eighteenth century. The small size of the doorway shows that this building has never been converted to a secondary use.

## CHURCH FARM, CLARE (TL 768 455)

Church Farmhouse is on the west side of the High Street, opposite the parish church. The dovecote is now 30 yards to the west, and slightly uphill from it, but it is not in its original position. In about 1890 an access road was formed through the farm to a new cemetery, and the dovecote was moved on rollers a few yards to the north to clear the route. There is no water nearby now, but probably there was a farm pond earlier. The building is 20 by 18 feet, 12 feet high to the eaves, with an off-centre door in the long side facing east towards the house (Figure 8). The walls are clad with lath and plaster, and weather-boards on the lower half. (This was often done in farmyards to protect the plaster from abrasion by farm animals). The plain-tiled roof is hipped with gablets to north and south, now boarded over.

Figure 8: The timber-framed dovecote of Church Farm, Clare, from the north-east.

38                    The Dovecotes of Suffolk

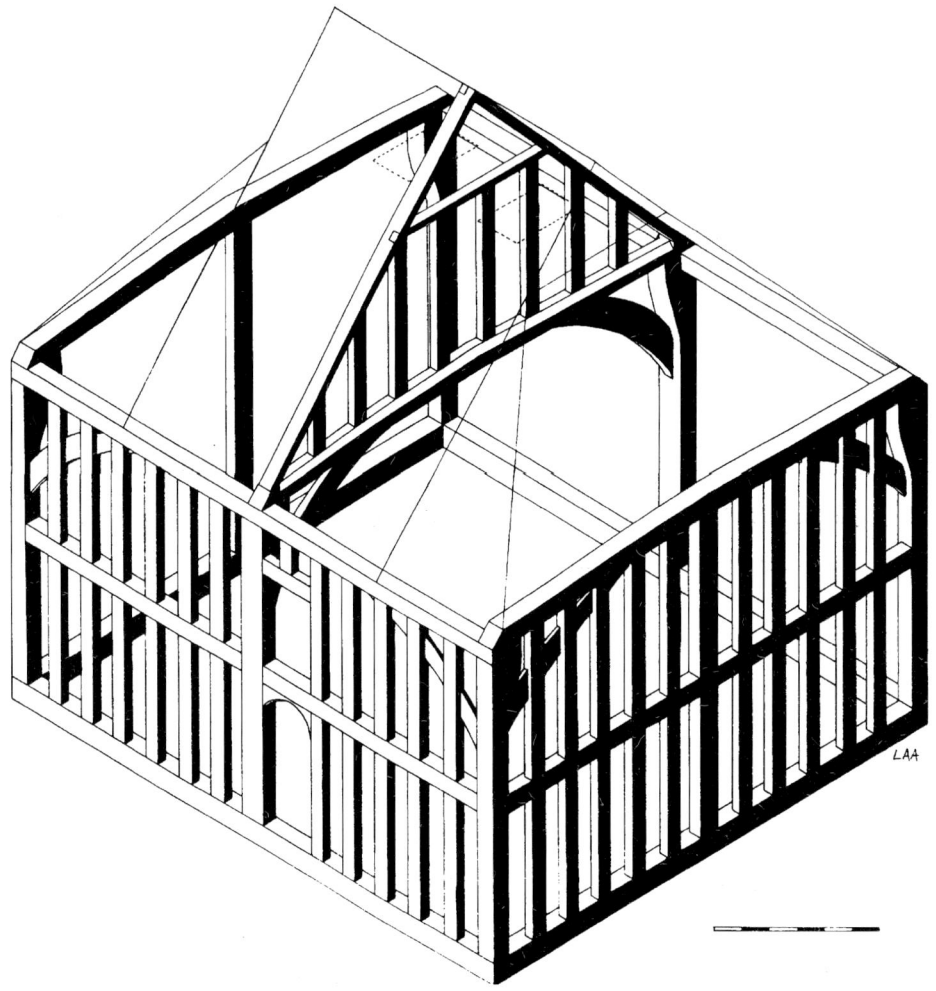

Figure 9: Isometric drawing of the dovecote of Church Farm, Clare, from the north-east.
(Scale in feet). By courtesy of Leigh Alston.

## *The timber frame:*
The original oak frame is intact, built in two equal bays, and most of the wattle and daub filling is present too, within the later cladding (Figure 9). The brick plinth is made of various re-used bricks in irregular bond, and dates only from the move to the present site.

## *The doorway and windows:*
The doorway against one central post has been enlarged, but remaining evidence shows that it was 2 feet 1 inch wide by less than 5 feet high (Figure 10). The window

# The Dovecotes of Suffolk

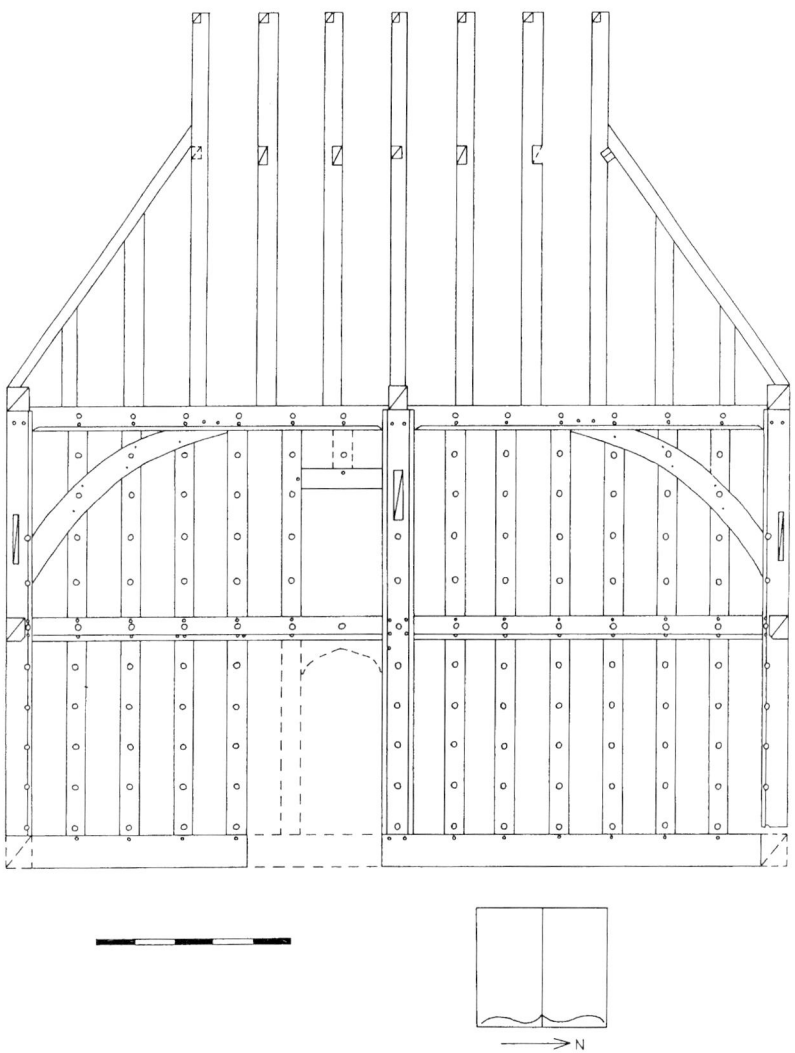

Figure 10: Vertical section of the dovecote of Church Farm, Clare, showing the east wall and doorway from the inside. (Scale in feet). Missing timbers are indicate by pecked lines.
By courtesy of Leigh Alston.

aperture above is of the same width, and 3 feet 2 inches high. It retains evidence of three former mullions, and is rebated on the outside, with traces of putty from former glazing. There is now a slatted window opposite, but this is a modern aperture which leaves the studding unaltered.

## *The former nest-boxes:*

All the nest-boxes have been removed, but a pattern of one-inch auger holes in the studs and posts indicates how they were formerly supported (Figure 10). The highest line of holes is in the wallplates on the long sides, continuing at this level on the studs of the short sides. Similar tiers of auger holes occur at one-foot centres below, sufficient for ten tiers of nest-boxes. If each box was one foot wide there would have been 64 boxes in each complete tier. Allowing for those omitted at the door and window there would have been about 624 nest-boxes in all.

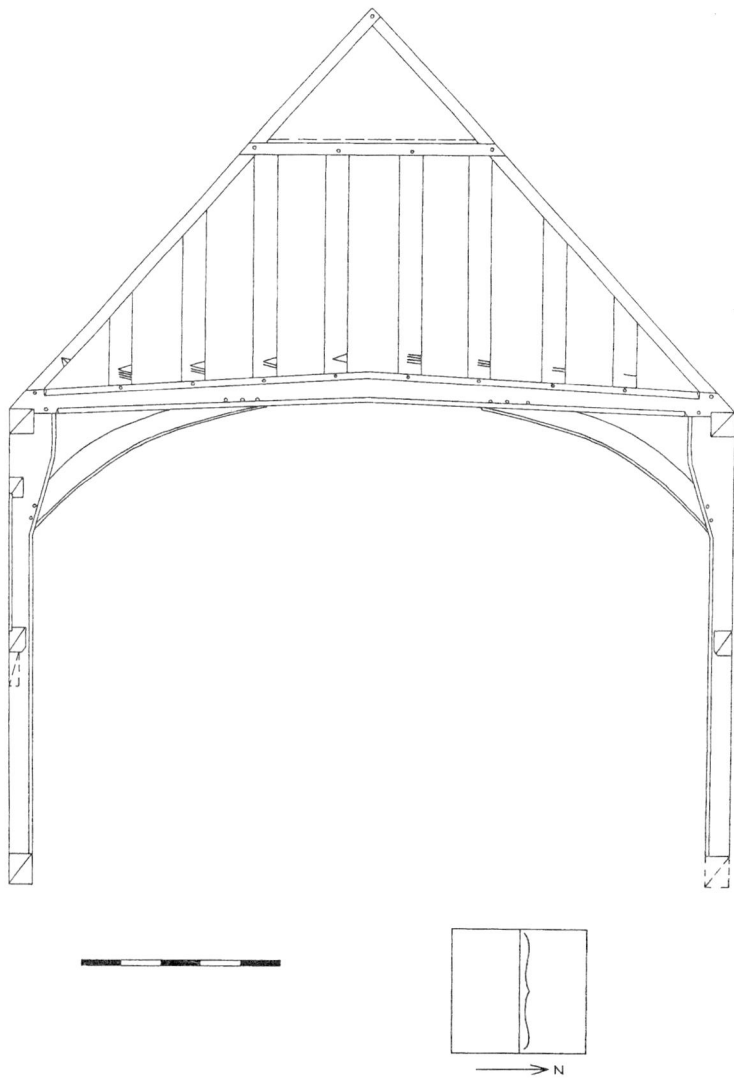

Figure 11: Cross section of the dovecote of Church Farm, Clare, showing the central truss from the north, and the scribed assembly marks. (Scale in feet). By courtesy of Leigh Alston.

## The roof:

All the original rafters are present, of horizontal section, with a collar lap-jointed across each full-length pair. There is a complete set of original sprockets.

There are two curious features in the roof construction. One is that above the cambered central tie-beam there is a line of original oak studs up to the collar, with fixings for wattle and daub (Figure 11). Evidently there was a partition above the tie-beam, which did not extend below it or above the collar. The other is that the upper arrises of the tie-beam are deeply chamfered (Figure 12a). It is normal to find that the lower arrises are chamfered, which was done mainly for visual reasons, and they are chamfered here. Chamfering the upper arrises served no visual purpose. Leigh Alston points out that when the daub filling between the studs was complete the chamfering would have left no horizontal surface on which a bird of prey might rest.

A flight platform has been built on the collars which span the roof, with two apertures for the pigeons to pass through. This is similar to early features elsewhere, but here the boards are of machine-sawn softwood, which would not have been available before the late nineteenth century; evidently it replaces earlier boards which were decayed. The present boards have decayed too, but in the north bay the aperture is still measurable, 2 feet 3 inches by 1 foot 9 inches; the latter dimension is determined by the interval between the collars. On each side of these holes the sides of the collars have been cut away at a steep angle to accept a funnel-shaped pipe (Figure 12b), similar to those already reported at Great Thurlow and 'South Suffolk A'. The pipes have not survived; nothing remains except some hand-made nails where the boards were attached to the collars. Two collars above the south bay retain similar evidence of a former pipe there. It was not unusual to have more than one pipe (in 1698 Roger North designed his dovecote for four pipes).[53] Here it was necessary to have two pipes because of the closed partition reported above.

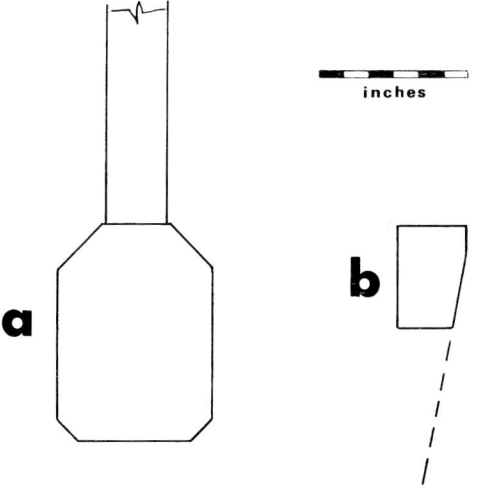

Figure 12: Sections of timbers at Church Farm, Clare:
(a) the central tiebeam and studding above, showing how the upper arrises are deeply chamfered to avoid leaving any perching place for birds of prey,
(b) a collar cut away at a steep angle to attach a pipe.

## Dating:

This is a characteristically sixteenth-century building. Alston reports that at that time Church Farm was owned by the wealthy Fenn family of clothiers. In 1544 John Fenn was one of three leading townsmen who leased the Clare manorial lands from the Crown. One suggestion is that he may have celebrated his new status by building a dovecote. If the window was indeed glazed from the outset this would be an early (though not impossible) date, for at that time glass was used only in the best windows of the richer houses. It became more easily and cheaply available after 1575.[54] A construction date in the late sixteenth century may be more likely.

## Later developments:

A floor of softwood joists and planks has been inserted 8 feet above ground. Below it the walls are lined with softwood boarding for housing geese. A vertical ladder rises through a trap in the north-east corner. Above the floor the framing and daub filling are complete and visible. On the inserted floor, among a large amount of broken clay debris, Alston found a complete *clay bat* 12 x 8 inches, with the upper surface deeply dished. Clay bats were made of chalky clay sub-soil mixed with water and straw, formed in wooden moulds and then allowed to dry, in the same way as the *clay lumps* which became common in nineteenth-century buildings in East Anglia. (Current research indicates that clay bats were used in dovecotes before clay lumps were used structurally).[55] In 1785 Daniel Girton wrote: 'It is necessary to have a small cavity sunk at the bottom of the coves, to prevent the eggs from rolling aside, for, though the pigeon may sit well in her nest, if this accident happens they will certainly be spoiled'.[56] If these bats were made to fit the original nest-boxes they supply the missing dimensions, but it is possible that they were made for a later arrangement in which the upper storey became a pigeon-loft, with new nest-boxes. The interior is lime-washed up to eaves level.

## Additional information:

The studs above the central tie-beam display a clear series of scribed *assembly marks*; they are on the north side, near the bottom, and number from I to VIII, west to east (Figure 11). It is likely that the studs of the outer walls were scribed in the same way, but if so the evidence is concealed by the later cladding.


## REDCASTLE FARM, PAKENHAM (TL 902 694)

The farmhouse is situated on high land two miles north-west of the parish church. It is of fifteenth or early sixteenth-century origin, with a parlour cross-wing built about 1600, enclosed on three sides by a moat. The dovecote is 15 yards east of it, just inside the moat, and is now used as a garden shed (Figure 13). It is 16 feet square, 11 feet high to the eaves, with a central door facing west-south-west (visible from the house); the only window is modern. The plain-tiled roof is hipped, with large gablets to north and south, now boarded over.

Plate 1a (top): The original roof and flight platform at 'South Suffolk A', with the funnel-shaped pipe through which the pigeons ascended and descended. Two of the boards have dropped away.
Plate 1b (bottom): The inner structure of nest-boxes at Chapel Farm, Gipping. The buff-coloured bricks are unfired, the red bricks are fully fired. Most of the nest-boxes have been blocked with bricks and plastered over for a later use, but one at left is shown still open. The lower part of the structure has been blacked and repaired with modern bricks.

Plate 2: The east elevation of the pigeon-tower at Stoke College, Stoke-by-Clare, showing the decorative design formed in flared headers - a portcullis, a letter I, and a bishop's mitre lying on its side.

Figure 13: The timber-framed dovecote of Redcastle Farm, Pakenham, from the west.

### *The timber frame:*

The oak frame has jowled corner posts, studded walls and minimal bracing (Figure 14). The present cladding is of cement render on modern lathing, but the absence of wattle fixings shows that it had external cladding from the outset. The walls have been lime-washed inside; this is relatively modern, for it covers some softwood insertions.

### *The brick plinth:*

The groundsills are raised a foot above present ground level on masonry plinths. In the south wall the plinth has been rebuilt with flint and brick rubble, but the remainder appears to be original, composed of red bricks 9¼ x 4¼ x 2 inches laid in lime mortar in *stretcher bond*. The present floor is of concrete.

### *The doorway:*

The doorway has been enlarged, but the spacing of the studs and the jointing

Figure 14: Isometric drawing of the dovecote of Redcastle Farm, Pakenham, from the north-west. (Scale in feet and metres). By courtesy of Leigh Alston.

for the former lintel show that originally it was only 4½ feet high by 2½ feet wide. The present door is of rebated softwood boards with hardwood ledges and braces, probably of the nineteenth century. The framing is fully exposed inside; there is no evidence of an original framed window, but a pattern of nails and staining on the studs shows that at some time there has been a window over the door, as drawn.

### *The former nest-boxes:*

The nest-boxes have been removed; a pattern of auger holes in the corner posts and studs (some still containing broken pegs) indicates how they were supported (Figure 15). The holes are of 1⅛ inch diameter, at approximately 14-inch vertical intervals. The nest-boxes are assumed to have been of sawn boards.

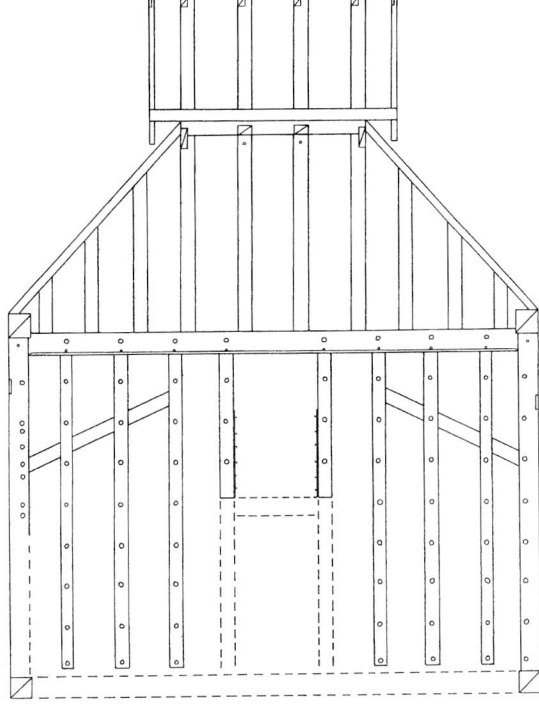

Figure 15: Vertical section of the dovecote of Redcastle Farm, Pakenham, showing the west wall and former doorway from inside. Missing timbers are indicated by pecked lines. (Scale in feet and half-metres.) By courtesy of Leigh Alston.

There is no evidence of their width, but on the assumption that they were 12 inches wide there is sufficient space for up to 450 nest-boxes.

*The roof:*

The roof is of oak rafters of horizontal section, with clasped purlins supporting collars which form the gablets. Resting across these purlins is a flight platform of massive oak planks, up to 3 inches thick, with a hole 1 foot 10 inches square in the middle,

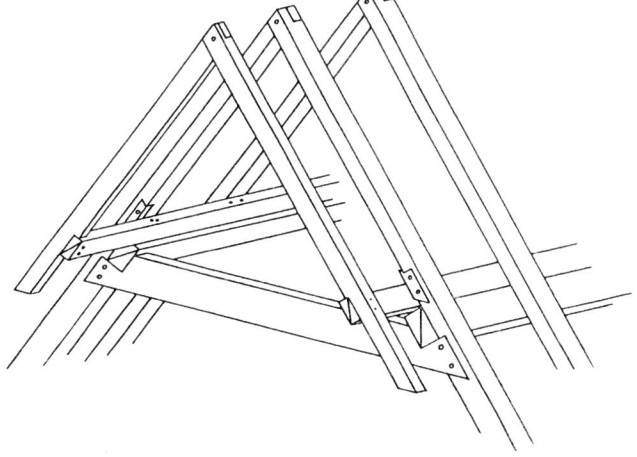

Figure 16: Detail of the dovecote of Redcastle Farm, Pakenham, showing how extra purlins were fitted after the original construction to provide shelter over the open gablets. By courtesy of Leigh Alston.

through which the pigeons ascended and descended. Additionally, two oak joists of horizontal section have been nailed across the purlins, and these outline the square hole. If there was formerly a pipe descending from this frame, all that remains is a series of nail holes round the edges. There is evidence of an early alteration to the roof, for extra purlins have been nailed over the main purlins, projecting to form triangular hoods which sheltered the open gablets from rain (Figure 16).

*Dating:*
The minimal bracing of the timber frame is typical of the early seventeenth century. The absence of wattle fixings, suggesting that it was clad from the outset with lath and plaster, tends to confirm this dating. One might hazard a guess that about the end of the sixteenth century the occupier achieved some prosperity and freehold status, and expressed it first by enlarging the parlour end of his house, and later by building a dovecote. As stated in Chapter 2, a change in the common law in 1619 enabled freeholders to build dovecotes, which earlier had been a privilege reserved to lords of manors and parish priests.

*Additional information:*
The farmer reports that traditionally the farm comprised about 200 acres. The number of nest-boxes deduced from the fixing holes seems proportionate to that area.

## CHAPEL FARM, GIPPING (TM 070 635)
Chapel Farm is a house of sixteenth-century origin, opposite the Chapel of St. Nicholas. The dovecote is 140 yards west of the farmhouse, in the corner of a field. It is octagonal in plan, each face 8½ feet across, 11 feet high to the eaves (Figure 17). The door faces east (towards the house), and there are high unglazed windows to north and south. When inspected in 1995 the building was dilapidated. There were holes in the plain-tiled pyramidal roof, and the louver had disintegrated.

*The timber frame:*
The structure is wholly of oak; the carpentry is designed and executed to a high standard. The angle posts are carefully cut to an hexagonal section to conform with the plan, tenoned and pegged to the groundsill and wallplates. Girts at mid-height are tenoned and pegged to them. Above and below each girt are three studs, tenoned at top and bottom, but only the middle studs are pegged. There is no bracing.

*The cladding:*
The cladding is unusual. Riven oak laths are nailed horizontally outside the posts and studs at 6-inch intervals; closely-spaced vertical laths are nailed to the inside of these laths. Daub consisting of yellowish clay mixed with straw is laid on each side, to the depth of the studding on the inside, and to a depth of 2 inches on the outside, making a total thickness of 5½ inches. The outside is painted with gas tar, a useful waterproofing material which became available from the early nineteenth century.

Figure 17: The timber-framed dovecote of Chapel Farm, Gipping, from the east. Despite its dilapidated condition most of the original cladding of clay daub has survived, coated with gas tar in the nineteenth century. The brickwork at right is a late repair.

### *The door and windows:*

The door frame is 5 feet 7 inches high by 3 feet wide, with the sill raised high above ground level. There is a ledged and boarded door on strap hinges which may be original, but is certainly early, with a large staple for a wooden bolt or draw-bar on the inside. The unglazed windows are 3 feet square, fitted with horizontal hardwood slats to keep out birds of prey.

### *The inner brick structure:*

A separate octagonal structure forms the nest-boxes, partly of fired bricks of the normal red colour, and partly of unfired bricks, which are buff in colour (Plate

1b, opposite page 42). These are not clay lumps for they do not contain straw or any other fibre, and the clay is sieved and *pugged* to a fine consistency, exactly the same as in the fired bricks. A solid plinth of fired bricks stands 2 feet high, and on this the nest-boxes are built. The front and sides of each box are formed of three courses of unfired bricks, the back and floor of fired bricks, so that the whole structure has a hard outer skin of fired bricks. All joints are of lime mortar. Each box is 8¾ inches high by 15 inches long by 8 inches front to back, with an entrance hole 6 inches high by 6½ inches wide. The floor of each tier projects to form an alighting ledge 2½ inches wide. There are ten tiers, with 32 nest-boxes in each complete tier. Allowing for those omitted at the door and windows there are 289 nest-boxes in this structure.

### The roof:

The roof is formed by principal rafters rising from the angles of the octagonal wallplate to an octagonal frame which formerly supported a louver. Joggled in-pitch purlins are tenoned and pegged to the principals. Common rafters of horizontal section overlay the purlins in parallel formation, nailed where they abut on the principals. Enough remains of the roof to show that the hips were leaded, and the louver had a leaded pyramidal roof. A full set of shaped sprockets survives *in situ*.

### Dating:

The Listed Building description of 1988 attributes the dovecote to the early eighteenth century, but further examination shows that there are two distinct phases of building. The timber frame is characteristic of the seventeenth century; the rafters of horizontal section and the daub cladding suggests that it was early rather than late within that century. The nest-boxes of that time have been removed, and there is no accessible evidence to show how they were formed. (There would have been more of them, perhaps nearer 400, for the space available was that much larger before the inner structure of bricks was intruded). The original building was easily penetrated by brown rats, and so would have become useless as a dovecote when they spread to this district in the eighteenth century. Evidently the inner structure of brickwork was built as a response to that hazard - using hard fired bricks to form the base and a protective outer casing, and softer unfired bricks on the inside where protection was not needed. Firing always made up a large part of the cost of bricks, so omitting that process allowed the builder to make a substantial economy. Another factor came into play from 1784, when fired bricks were taxed, adding about 15% to their cost; unfired clay products were exempt. The inner structure was built after the middle of the eighteenth century, and probably after 1784. It enabled an otherwise indefensible building to be brought back into use as a working dovecote.

### The former potence:

There is evidence that a *potence* or revolving ladder has been used. A beam has been set across the brick structure, near the top. In the middle an iron plate is attached to the *soffit*, with a round hole to take the upper pivot. Nothing else of it survives.

### Later use:
Nearly all the nest-boxes have been blocked later, for some use of the building in which all these cavities comprised a nuisance. To a height of 5½ feet they have been faced with fired bricks, and the vertical surface has been tarred. Above that level they have been blocked with bricks and roughly plastered over. In addition there are later repairs with flettons, which did not become available until the 1890s. It is now used as a log store.

### Additional information:
The unfired bricks are 9½ x 4½ x 2½ inches. The fired bricks are marginally smaller, corresponding with the normal shrinkage which occurs during firing. The extrusions from the mortar joints are left protruding on the outside. The brickwork over the door aperture is supported on two oak lintels of 8 x 5 inch horizontal section, with similar lintels over the window apertures. It is to be hoped that eventually it will be possible to date this building by *dendrochronology*; the posts and principal rafters would date the original timber-framed structure, and the lintels would date the inner structure of nest-boxes.

A medieval dovecote in this parish is mentioned in Dr. Hoppitt's Appendix.

## ASPALL HALL, ASPALL (TM 168 647)
This dovecote (or what there is of it) is isolated 500 yards south-west of Aspall Hall, in the corner of an apple orchard. It stands on a flat-topped mound 8 feet above ground level, surrounded by a ring ditch. The photograph was taken in 1947 (Figure 18); all that remains now is a roofless ruin, overgrown by ivy, wild roses, elder and nettles - difficult to approach or even to see except in winter when the vegetation has died down. It is not Listed.

Although captioned as a dovecote in 1947, it had already been converted to modern use (as will be described later). It is hexagonal, 8½ feet across each face, 16 feet high to the eaves, with a doorway to the south-east and windows to east and north-west. The pyramidal roof was formerly clad with pantiles, with galvanized iron hips. The rustic bridge shown in the photograph has gone.

### The timber frame:
The angle posts are of oak, approximately 8 x 6 inches, but cut to a pentagonal section in accordance with the plan of the building. They are mounted on oak groundsills of 6 x 3½ inch vertical section on a brick plinth, and are connected at a height of 9 feet by oak girts 5 inches square. These are the timbers which show as black in the photograph. Each post has been supported on the inside by a later earth-fast post standing 3½ feet high, bolted to it. The wallplates and roof timbers have gone. The door-frame is original, 6 feet high by 2½ feet wide, with chamfered oak jambs. The windows had softwood frames, and were later insertions. There is a softwood beam across the full span, supported on bolted brackets, also modern. Most of the original studs have been replaced in softwood.

Figure 18: The timber-framed dovecote of Aspall Hall, Aspall, from the north-east, standing on a man-made mound and surrounded by a ring ditch. Photograph by H. N. Collinson, published in 'East Anglian Magazine', December 1947. By courtesy of Suffolk Record Office.

### *The brick plinth:*

The plinth stands six courses above ground. The bricks are hand-made, 9 x 4½ x 2½ inches, laid in lime mortar in Flemish bond. At a later date the inside has been treated with cement-sand render.

### *The cladding:*

Some areas of original cladding remain, consisting of riven oak laths fixed horizontally to the frame with hand-made nails, coated with clay daub containing small stones. Apparently this cladding was intact when photographed in 1947, but with the destruction of the roof and the passage of fifty years much of it has weathered away.

### *The filling material:*

Distinct from the cladding is a kind of filling material which has not been noted elsewhere. It comprises slabs of chalky clay, sieved and ground to a fine texture, and left unfired. These are not clay bats in the usual sense because they do not contain straw or other fibrous material. They appear to have been moulded specially to fill this frame, for they fit accurately against the pentagonal posts, and they have smooth plane surfaces and sharp arrises. They are all 3½ inches thick, but they vary in size from 15 x 7 inches to 21 x 13 inches according to their positions in the frame. The ¼-inch mortar joints are of the same clay. These clay slabs probably represent the backs of the former nest-boxes; the floors, sides and fronts would have been removed when the building was converted to modern use.

### *Dating:*

This is a difficult building to date because so much of it has gone. The style of the carpentry suggests the early eighteenth century. There are some parallels with the earlier octagonal dovecote at Gipping.

### *The defended site:*

This is the only example known in Suffolk of a dovecote built on an artificial mound surrounded by a ditch. It is not known whether the mound was first built for a mill, or whether it was raised specially for the dovecote. It was always desirable to mount a dovecote high, because pigeons are instinctively attracted to high places, but the more significant feature here is the ditch. It is of V-section, 10 feet wide, and was formerly connected to the nearby boundary ditch, functioning as a water-filled moat. More recently the boundary ditch has been deepened, leaving the ring ditch dry.

All animals can swim to avoid drowning, but polecats and cats do not take to water willingly, so surrounding a dovecote with a moat was a highly effective defence against these predators. In 1577 Barnaby Googe wrote: 'There be some that build their Dovehouses upon pillares in the middest of some Pond, or great water, both because they [the pigeons] delight in water, and because they will have them safe from Vermine'. (In the Appendix Dr. Hoppitt reports documentary evidence of this practice in 1385: 'the dovecote within the great pond below the castle at Framlingham'). John Worlidge confirmed the value of a water barrier in 1669: 'To prevent Pollcats or such like from destroying your Pigeon-house, be sure if you can to erect it, where you may have a Ditch or Channel of Water to run round it, and it will keep those vermin from making their Burroughs under the ground'.[57] Edward Martin has identified other sites in Suffolk where a small moat - too small to surround a house or cattle pen - appears to have been made to protect a dovecote, but it is difficult to prove when the building has gone.

Brown rats swim readily and spread most easily along water-courses, so the moat which was intended as a defence against the earlier predators became an easy access route for the later ones. When this became apparent the moats round dovecotes were filled in. There are other sites where one suspects that the dovecote was formerly defended by a moat, but only archaeological excavation would provide sufficient proof.

### *Earlier descriptions:*

In this case we are fortunate in being able to draw upon the recollection of Mrs. P. M. Guild, who was born in 1902 and who lived at Aspall Hall. She remembers that the dovecote had what she describes as 'a pole with shelves on it' when she was a child, which may have been the axis of a potence, fitted with feeding platforms.

At that time two tiers of nest-boxes remained *in situ*, although the building had already ceased to house pigeons. Her father converted it to an apple store, used with the adjacent orchard. She recalls that later it was used by Boy Scouts. At some stage an architect drew up plans for a residential conversion, but the project was abandoned. She also remembers a lady collecting pictures of dovecotes in Norfolk and Suffolk in order to depict them in needlework. It would be interesting to know where those representations are now.

Claude Messent reported in 1936 that the exterior was then roughcast, with small pieces of green bottle glass set in the surface, some of which outlined the numerals 1919 - which determines only the age of the render. The roof was still pantiled and hipped with metal, and the original door was present.[58]

## *Additional information:*

An estate map of 1621 appears to show a dovecote on one side of the farm courtyard of Aspall Hall, next to a pond.[59]

**(One more timber-framed building will be described, a pigeon-loft integrated with a kennels range at East Bergholt, but as it is the latest in date of the whole series it will be left until the end of Chapter 7).**

# Chapter 6
# Brick dovecotes

**STOKE COLLEGE, STOKE-BY-CLARE** (TL 740 433)

Stoke College is an independent boarding school on the site of an alien priory of Bec Abbey, which was dissolved and established as a college of secular canons in 1415. The pigeon-tower is beside the main entrance, and 200 yards from the College itself (front cover). The River Stour is 120 yards to the south-west. This has always been a dual-purpose building, comprising a pigeon-loft over an open lower storey. It is 16 feet square and 21 feet high to the eaves. At the north-west corner the brickwork was originally bonded into a brick wall 8 feet high extending to the west, but only a fragment of that wall remains. A later wall has been connected at the same point, which forms the present park boundary.

This building is sometimes wrongly identified as a gatehouse, but examination of the lower storey shows that it was designed as a pavilion, a place where the canons could obtain shelter from the rain or sun while out in the grounds (Figure 19). On the south and west sides there are wide archways facing into the park, and on the north and east sides there are solid walls which formerly had small ornamental apertures facing the public road and access road. The pigeon-loft has one window facing south. The plain-tiled roof is hipped, with large gablets to east and west.

Figure 19: The pigeon-tower at Stoke College, Stoke-by-Clare, from the south-south-west.

### The brickwork:

The bricks are 9 to 9½ inches long x 4½ inches x 2⅛ inches, laid in lime mortar in English bond, four courses rising 10½ inches. They are of unsieved clay, wood-fired to various colours - orange-red, red and plum, and some are over-fired to a dark blue colour and vitrified. On the north and east elevations these flared bricks are used decoratively, as will be described.

### The pigeon-loft:

Because this work is about dovecotes the pigeon-loft should be considered first. The south window is unglazed, 3 feet 6 inches high by 1 foot 7 inches wide, with a chamfered three-centred arch. The original floor structure remains unaltered, consisting of oak joists 7 inches square and 7 inches apart, set in the north and south walls 8 feet above ground, with a framed trap in the south-east corner. The floor-boards have been renewed.

### The nest-holes:

The nest-holes are integrated with the structural brickwork, making a wall just over 2 feet thick (Figure 20). There are 13 tiers of nest-holes, each 6½ inches high, 9 inches wide, 10½ inches from front to back, with an entrance hole 6 inches high by 4¼ inches wide at one side. *Headers* project to form an alighting ledge 2½ inches wide to each tier. Each complete tier has 45 nest-holes, so allowing for those omitted at the window there were 565 in all. Three below the window have been blocked by modern repairs, and a few nest-holes and ledges have been damaged, but otherwise they remain in good order. The whole interior has been lime-washed, although little of it remains visible except inside the nest-holes.

Figure 20: Inside the pigeon-tower at Stoke College, Stoke-by-Clare, showing the western and northern nest-holes.

### The window:

The internally splayed window is slightly off-centre, its position apparently determined by the arrangement of nest-holes. Four tiers of nest-holes are carried over it on an oak lintel which projects to form a continuation of the adjacent alighting ledges. Surprisingly, there is no trace of an iron grid to protect the window against birds of prey, nor a rebate for anything else. However, close examination reveals some less conspicuous evidence that a protective barrier was fitted. All the brickwork inside the loft is scored by small vertical grooves made by the pigeons, as is usually found in dovecotes which have been in use for a long time. These grooves occur in the window splays to within 4 inches of the inner surface of the wall, but not outside that line. This suggests that a wooden lattice was fastened across the splays by nails.

### *The lower storey:*

The archways are 7½ feet wide by 7 feet high. The south arch is four-centred, the west arch more Tudor in shape (that is, consisting of two short-radius curves and two straight lines). Chamfers inside and outside are formed with bricks cut and rubbed to shape. There are remains of lime plaster and red paint on the soffits, probably original. The arches are formed of rectangular bricks with tapered mortar joints, except for a wedge-shaped brick at the apex of each. At the short-radius curves the mortar joints are packed with tiles, and in the west arch also with oyster shells. These details suggest that originally all the brickwork was covered by a thin layer of lime plaster, and painted. Timothy Easton has shown that this decorative treatment of brickwork was common at all periods.[60]

The north and east walls each have two deep niches with corbelled heads, arranged slightly asymmetrically. Each niche was built with an aperture finely formed with cut bricks to the outline of a Calvary cross (that is, a Latin cross on a stepped pedestal). The apertures have been blocked at an early period with similar bricks and lime mortar, but the outlines are still clearly visible from outside.

### *The roof:*

All the original oak rafters (of horizontal section) are present, with attached trims of modern softwood, and modern lathing. The north and south wallplates project beyond the brickwork. The others are tenoned into them at the same level. Two collars are halved across rafters of the north and south pitches, two cross-pieces are laid over them to form the bases of the gablets, and the hip rafters are pegged to these cross-pieces. Three pairs of short rafters rise from the collars to form the high roof. The flight platform is missing, probably taken away when the roof was re-tiled. Most of the sprockets are original.

### *The decorative brickwork:*

The most striking features of the exterior are the highly accomplished designs formed with *flared* bricks in the north and east elevations (front cover and Plate 2, opposite page 43). High on each wall is a depiction of a portcullis with a loosely arranged chain and ring to each side. Below it is a letter I with exaggerated serifs, and what appears to be a bishop's mitre lying on its side. Traces of diaper patterns are visible below the eaves. It is probably the finest piece of decorative brickwork in a county which is richly endowed with fine brickwork, and emphasizes the high status of the building. I am grateful to Edward Martin for the following interpretation:

The portcullis was first used as an heraldic device by the Beauforts, Dukes of Somerset, and was adopted as a royal device by King Henry VII, whose mother was a Beaufort. The college of priests was under the patronage of the Queens of England, and so was entitled to display this royal device. The letter I stands for St. John the Baptist, to whom the college was dedicated. The mitre can be seen as such, or as a Lombardic letter E, for it ingeniously represents both. It refers to Richard Edenham, who was dean of the college from 1470 to 1493, and also Bishop of Bangor - the only dean who was also a bishop. It is not significant that on the east elevation the E is reversed, for there are many parallels. For instance, in early inscriptions in arabic numerals the digits are often reversed.

### *Dating:*

Were it not for these devices it would be difficult to date the building beyond saying that it is characteristically Tudor, but it can be firmly attributed to Richard Edenham, and to that period of his tenure which followed the accession of Henry VII, between 1485 and 1493. Few buildings of the period can be dated so satisfactorily. Perhaps the most surprising fact about this building is that it was Listed as Grade II* in 1974, and still remains at that grade. It is of such high quality, and so little altered, as to be of national importance, and certainly merits Grade I.

### *Additional information:*

There are two solid courses above the floor, and three more above the nest-holes and below the wallplates. Each tier of nest-holes occupies four courses. All the nest-holes have off-set entrances, but the arrangement is irregular at window level. In the west wall the entrances are to the right. From the north-west corner to the east side of the window all the entrances are to the left. In the short section of the south wall to the west of the window they are to left and right in alternate tiers. One gets the impression that the bricklayer built the first four tiers of nest-holes without working out how this pattern would be affected by the window (perhaps concentrating more on incorporating the decorative brickwork of the exterior). This left the window slightly off-centre, and disrupted the regular pattern of nest-boxes from there onwards.

## FLIXTON HALL, FLIXTON BY BUNGAY (TM 304 859)

Flixton Hall was built about 1618; the dovecote is contemporary with it, and was then a free-standing building. The Hall was destroyed by fire in 1846, but its appearance is known from contemporary engravings. A large country house in Jacobean style was built 1888-92, and this in turn was demolished in 1952. In the nineteenth century high walls were built out to meet the dovecote, so it now forms the north-east angle of a walled courtyard. Apart from the yard buildings, a laundry, and some low walls and terrraces nothing else remains of what was formerly a major establishment.

The dovecote is cylindrical, 22 feet in external diameter, and 22 feet high to the eaves (Plate 3a, opposite page 58). The nineteenth-century walls which abut to south and west are 15 feet high with crenellated heads. The door is to the north-west, with a window above, and there are three other windows. The conical roof has been rebuilt in the nineteenth century, and is clad with machine-made red clay tiles, with a damaged octagonal lantern of the same period.

*The brickwork:*
The bricks are 9½ x 4½ x 2¼ inches, laid in lime mortar in English bond, four courses rising 10¾ inches. They are of unsieved clay, *clamp-fired*, fairly uniform in size, mostly red, but some are burned to a plum colour. The structural wall is 13 inches thick, on a plinth 3 feet above present ground level with a weathered set-back 2½ inches wide. There is neither string course nor cornice.

*The door and windows:*
The chamfered oak door-frame is original, 4 feet 4 inches high by 2 feet 2 inches wide, jointed with pegs of ¾-inch diameter, but the adjacent brickwork has been altered. The sill is now 2 feet above ground level, with bricks of eighteenth-century quality below it. Evidently the doorway was raised in the eighteenth century to secure it against brown rats. They can jump up to two feet, but probably at that time the ground level was kept lower. The door itself is nineteenth-century, of pine, the upper half consisting of a grill of vertical ½-inch iron rods ¾-inch apart, opening inwards. Inside there are three steps descending to the earth floor, which is substantially below the outside ground level.

The least altered window is to the north-east, 13 feet above ground. It has an oak frame with a central mullion, vertical saddle bars of diamond section to each side, and leaded glazing. Inside there is a grill of vertical iron bars 2 inches apart, for protection against birds of prey when the glass failed (for seventeenth-century glass was fragile, and needed repairs frequently). The window above the door is similar in design, but it has been rebuilt with a pine frame at some time, probably when the doorway was altered in the eighteenth century. Another original window aperture to the south-west (facing into the later stable yard), also 13 feet above ground, has been boarded over. An unglazed window to the south-east, 9 feet above ground, appears to be a nineteenth-century insertion; it is protected by horizontal hardwood slats. At some date a doorway has been roughly broken through from the stable yard, and has been blocked again with re-used bricks.

*The nest-boxes:*
The brick nest-boxes form an independent structure inside the structural wall. The lowest tier is 2 feet above internal ground level (Figure 21). Each box is 7½ inches high, 10 inches wide, and 10 inches from front to back, with an entrance hole 5½ inches high by 4 inches wide at the right side. The lowest three tiers have continuous alighting ledges of headers which project 1½ inches, but above that each entrance hole has a separate alighting step. There are 23 tiers, and 54 nest-boxes to each complete tier. Allowing for those omitted at the door and windows

Figure 21: Inside the dovecote of Flixton Hall, Flixton by Bungay. Three steps descend from the raised doorway into the interior. Note the unusual ladder.

this makes a total of 1,178. The brick floor of each box is mounted on a thin board which was used to support the bricks while the lime mortar was setting.

The whole interior has been lime-washed many times, including the insides of the nest-boxes.

### *The potence:*

In the middle of the floor is a cylindrical plinth of nineteenth-century brickwork, 7 feet in diameter and 3½ feet high, supporting the lower bearing of a potence.

Plate 3a (top): The early seventeenth-century dovecote of Flixton Hall, Flixton by Bungay, from the west-north-west.
Plate 3b (bottom): The ruined dovecote at Goswold Hall, Thrandeston, from the south-east, and Goswold Hall beyond.

Plate 4: At Flixton Hall, showing the potence with its unusual ladder and large feeding platform.

The octagonal axis is older than the remainder, probably of eighteenth-century origin. The two bracketed arms of machine-sawn oak with iron fastenings evidently represent a nineteenth-century rebuild. The ladder they support is basic, and perhaps indicates the difficulty of obtaining a sufficiently long ladder of normal form (Figure 21 and Plate 4, opposite page 59). A single inclined bearer of oak 5½ by 3 inches has rungs 9 inches long and 9 inches apart projecting on alternate sides, except the lowest two which project on both sides. Many of the rungs are decayed or unreliable. The upper bearing of the axis is at the crossing of two beams set in the top of the brickwork.

## *The feeding platform:*

An octagonal platform 10 feet across is attached to the axis 13 feet above ground, supported on eight tangential arms each bracketed from the axis (Plate 4). The whole structure is of softwood fastened with nails - a typically nineteenth-century construction. Smaller feeding platforms have been recorded at Dunster in Somerset, Castle Bytham in Lincolnshire, and at Corby Castle in Cumberland, but nothing of this size is known elsewhere. Evidently it was built so that the pigeon-keeper could put out containers of food and water inside the building, at a height which was equally accessible to all the pigeons. In earlier periods they would have been fed outside, but this indicates a change of practice in the early nineteenth century.

## *The roof and lantern:*

The nineteenth-century roof is simply constructed. Eight principal rafters rise from the wall-head (which could not be inspected), and curved purlins are jointed to them with through-tenons and face pegs. Everything else is nailed. There are eight common rafters below each purlin, five above, converging to form a conical curve; some are of re-used timber. They rise to a round frame which supports the lantern. Within it a smaller round frame forms the upper edge of a pipe, estimated to be 3 - 4 feet in diameter. A few of the vertical boards remain *in situ*, but most have fallen away as the nails decayed. The depth could not be measured, but appears to be the usual 2½ feet.

The nineteenth-century octagonal lantern has a round arch in each face. When inspected it was in poor condition. One corner post was missing, and only fragments of the glazing bars remained, but as far as could be determined it had had six glazed lights on each side, with a slot below through which the pigeons entered. Its roof is pyramidal, clad with lead. On it is mounted a wrought iron weather-vane with a horseshoe welded to the north arm; the pivoted arm represents a forked pennant.

By the number of nest-boxes this is the largest dovecote in the county (although a larger building at Eriswell which formerly contained more nest-boxes will be reported in Chapter 8). Evidently it served a major estate. The small doorway and central structure confirm that it has never been converted to a secondary purpose. Despite its age it remains in sound condition. Its importance is expressed by its Listing as Grade II*. One hopes that it will continue to be maintained.

*Additional information:*

Pevsner described this building as 'of red brick, and oddly enough, semicircular in plan', and this error was repeated by M. J. A. Beacham in 1990.[61] The illustrations show that it is fully cylindrical.

## GOSWOLD HALL, THRANDESTON (TM 125 755)

Goswold Hall is a moated farmhouse dating from the sixteenth century. The ruined dovecote is isolated in former parkland 130 yards to the east-south-east, near the main access road. It is rectangular, 14 feet by 12 feet, and 12 feet high to the eaves (Plate 3b, opposite page 58). The doorway was in the middle of the long side facing north (towards the access road), with a window above, and there are traces of another window in the east end. The building was already roofless when it was Listed as Grade II in 1987, and it has suffered further damage since. By 1996 it was effectively a ruin - but revealing its structure more clearly than when intact.

The end wall has a finely formed curvilinear gable, and a projecting eaves course following the profile; it is deeply cracked in two places. Above the doorway the brickwork has collapsed. Most of the south-east corner has fallen. Iron tie-bars in both directions and an inserted tie-beam show that the walls were spreading even before the roof was lost.

*The doorway and windows:*

The doorway was 4 feet 9 inches high by 2 feet 4 inches wide. The window immediately above was 1 foot 6 inches high by the same width. Only the decayed frame remains, with a mortise for a missing central mullion, and a vertical iron bar still *in situ* to each side. Of the former window in the east end (described as a small opening in the Listing notice of 1987) nothing remains except one jamb and the fallen lintel.

*The brickwork:*

The bricks are 9 x 4¼ x 2⅜ inches, laid in lime mortar, four courses rising 11½ inches. The bricks exposed on the outside are of finely sieved and pugged clay; most are red, but some are over-fired to blue. The bricks used internally are of markedly poorer quality, many insufficiently fired. Now that they are exposed to the weather they are eroding badly. Purpose-made bricks 18 inches long (but of the same width and depth as the others) form the floors of the nest-holes (illustrated on back cover). The pattern displayed outside is similar to Flemish bond, but with one irregular course in four.

*The nest-holes:*

The lowest tier of nest-holes is at ground level. Each is 9½ inches high, 15 inches wide and 8¼ inches from front to back, with an entrance hole 6 inches high by 5 inches wide. All those in the main tiers have the entrance at the right side, but in the gables they are necessarily irregular. Below each entrance a header projects to form an alighting step. When complete there were 12 tiers, with 26 in each complete tier, and 13 more in each gable. Allowing for those omitted at the door and windows there were about 310 in all. All the nest-holes have been lined with clay daub containing straw and chips of chalk, half an inch thick (back cover). The lining is so neatly done that it must have been executed as the brickwork rose. Some contemporary writers expressed the view that soft materials such as clay or plaster formed warmer and more comfortable nesting places than bare bricks, more attractive to the pigeons.[62] Evidently the owner of this dovecote shared that view.

## Dating:

The style and brickwork are characteristic of the late seventeenth century. When complete this was a building of high quality, executed with great skill. Neither the number of nest-holes nor the farmhouse itself suggest any great opulence. The high quality tells us more about the general level of building craftsmanship of the period than about the status of the owner.

## Additional information:

In 1936 Messent reported that the plain-tiled roof and both curvilinear gables were present, but that the walls were already cracked and tied with iron plates; he took this to be early settlement, caused by building on soft ground. He also described the louver: 'The louvred vent with the pigeon holes is in the centre of the ridge of the roof'; it was roofed with plain tiles. So few original louvers remain today that even this short note is helpful.[63]

Traces of lime-wash on the outside indicate that at some period the exterior has been whitened. This was done to attract and retain the pigeons. There is a large oak tree 5 yards to the east; measurement of the girth shows that it is about eighty years old; it would not have been present when the dovecote was in use.

## 'SOUTH SUFFOLK B'

As with one of the timber-framed dovecotes already reported, by the owner's request this dovecote will be described but not identified. Readers are asked to respect the owner's wishes, and not to try to visit it.

It is situated on a high site remote from any building or water. The dovecote is octagonal, each side of 7½ feet, and 17 feet high to the eaves (Figure 22). The door faces south, with a window above, and there is another window opposite. The plain-tiled roof is pyramidal, with a modern louver.

## The brickwork:

The bricks are 9 x 4¼ x 2½ inches, slightly irregular in depth, of sieved and pugged clay, laid in lime mortar in Flemish bond; four courses rise 12½ inches. A plinth six courses above ground forms a plain set-back of 1½ inches on the outside. Above the plinth the walls are 13½ inches thick. At the angles the bricks are cut and rubbed to shape. Below the eaves is a simple cornice formed of a continuous course projecting one inch, a course of separate headers forming *dentils,* and a continuous course projecting one inch beyond the dentils. The shallow segmental arches above the windows are formed of rectangular stretchers with tapered mortar joints.

Figure 22: The dovecote identified as 'South Suffolk B' from the south-east. The louver is modern.

## *The door and windows:*

The hollow-chamfered oak door-frame is original, the aperture 6 feet 6 inches by 2 feet 7 inches, with a limestone threshold and a modern door opening inwards. This is the largest unaltered doorway reported so far. The unglazed window immediately above is of the same width and 4 feet high; it now has widely-spaced modern laths set horizontally. The north window is similar.

## *The nest-boxes:*

On the inside is a brick platform 2 feet above ground, on which the nest-boxes are built. They are made of wattle and daub supported on battens of cleft oak 2½ x

1 inch which are built into the perpendicular joints of the brickwork, projecting 1 foot 8 inches (Plate 5, opposite page 68). The wattle is composed of coppice wood ½ to 1 inch in diameter and some cleft laths, bound to the battens with coarse twine. The daub is of chalky yellow clay with a high straw content. There are 12 tiers of nest-boxes, with 40 in each complete tier. Allowing for those omitted at the door and windows there were 456 in all, most of which are still in good order. Against the straight sides of the octagon the nest-boxes are 12½ inches high, 14 inches wide, and 11 inches from front to back; at the angles they are necessarily constricted. The entrance holes are 6½ inches high by 5 inches wide, all offset to the right. There is a continuous alighting ledge 2½ inches wide to each tier, formed in the same way.

These nest-boxes are evidently original, for the battens supporting them are integral with the brick structure. It is significant that in the lowest tier the brick platform which forms the floors has been coated with clay daub. As noted at Goswold Hall, some writers were advocating that pigeons 'liked' nests of the softer materials. How the interior was furnished depended mainly on what the owner or pigeon-keeper believed.

### *The potence:*

There is an incomplete potence, comprising an oak axis of octagonal section 7½ inches across mounted on a pivot in a similar timber set in the ground, standing 6 inches above it. In accordance with milling practice the ends are bound with iron hoops. Two beams of horizontal section span the top of the brick structure and are halved together. A length of hoop iron with a hole in the middle is nailed across the joint to form the upper pivot (Plate 5). Originally an inclined ladder was supported on three pine arms extending horizontally from the axis at heights of 2½, 7 and 16 feet above the floor, threaded through mortises and secured by wedges. Only the highest arm is intact, with the stub of a supporting bracket, and a triangular board at the end against which the ladder was secured. The other arms are decayed; about half of each remains. In addition a feeding platform was formerly present, at the same level as the lowest arm. It was supported on it, and on three smaller arms, but only decayed stubs remain.

The architectural design and the brickwork are characteristic of the second half of the eighteenth century. The high and remote position would have been attractive to pigeons - if inconvenient for their keeper. However, it is likely that its position was determined as much by aesthetic considerations as by functional ones, for evidently it formed a deliberately conspicuous feature in a planned landscape. Its unusual height, and the height of the combined door and window aperture, seem to be determined primarily by visual considerations.

### *Protection against brown rats:*

The high position would not have attracted brown rats, which spread mainly along water channels. They would have reached south Suffolk before this dovecote was built, but were mainly a nuisance on lower ground. The construction was fairly secure against them, with the possible exception of the door. The original door has not survived, but a sheet of iron across the lower part would have protected it sufficiently. The feeding platform would have been above the reach of brown rats. The nest-boxes, 2 feet above the present earth floor, are barely high enough for full protection, but perhaps the floor level was lower when the dovecote was in use than it is now.

*Additional information:*
   The brick platform on which the nest-boxes are built is raised on seven segmental arches, which was done mainly to economize on bricks. There are parallels elsewhere - for example, at Home Farm, Newton, Cambridgeshire. This platform has been lime-washed, but not the remainder.
   The roof and louver are modern, built by a local carpenter in the 1980s. The louver is octagonal, with sets of parallel inclined boards on each side, and a pyramidal roof. The inclined boards are closer together than they would have been originally, but pigeons are not encouraged to enter the building now.


## CROWFIELD HALL, CROWFIELD (TM 137 576)

Crowfield Hall was built for Sir Henry Harwood in 1727-8; it has been demolished. The roofless ruin of a dovecote now stands isolated, 100 yards north-east of the nearest building, a cottage converted from the former stable range. It is octagonal, 10 feet across each face, demolished to a height of 10 feet (Plate 6a, opposite page 69). An old photograph in the possession of the farmer shows it as of nearly twice that height. An aperture 8 feet wide by 8 feet high with a concrete lintel has been made in the north-east side, evidently on the site of the former door. There is no evidence of windows, which must have been above what is left of the walls. It is not Listed.

*The brickwork:*
   The bricks are $9\frac{1}{2}$ x $4\frac{1}{4}$ x $2\frac{1}{2}$ inches, laid in lime mortar in Flemish bond; four courses rise $11\frac{1}{2}$ inches. They are of unsieved clay with a sandy texture, hand-made and wood-fired, mainly red but with some over-fired to blue. On the outside is a plinth seven courses above ground, with a set-back of $2\frac{1}{2}$ inches. Although the building is octagonal externally, inside it is cylindrical, $22\frac{1}{2}$ feet in diameter.

*The former nest-boxes:*
   The nest-boxes have been demolished, but headers and broken half-bats project in vertical lines at intervals of 1 foot 4 inches; these are the stubs of radial partitions which formed the sides of the nest-boxes. A few horizontal laths survive between these projecting bricks, which indicate that the floors were of plaster on laths. Sufficient evidence remains to deduce that the nest-boxes were $10\frac{1}{2}$ inches high and 16 inches wide, depth from front to back unknown. There would have been 41 nest-boxes in a complete tier. Traces of 9 - 10 tiers remain, so taking the height from the old photograph one might reasonably suppose that there were 15 - 18 tiers originally. This would yield a total of 615 - 738, less those omitted at the door and window apertures.

## Additional information:

In 1976 Norman Scarfe mentioned a dovecote dated 1731 at Crowfield Hall; this may not have been the same building, for thirty years earlier Messent had described it as 'fast falling into decay and now nearly past repair'.[64] The inside has been cement-rendered to a height of 4½ feet for later use as housing for cattle. The outside is marked by numerous nail-holes, which shows that fruit has been grown against it. The farmer reports that his father remembered a potence *in situ*.

## POLSTEAD PONDS FARM, POLSTEAD (TL 990 382)

Polstead Ponds Farmhouse is a handsome Georgian brick house on the north side of the Box valley, with large ponds 30 yards to the south and south-east. The dovecote is 13 yards south-east of the house, with the farmyard extending beyond. The site declines steeply to the south. The dovecote is rectangular, 15½ feet by 14 feet, 12½ feet high to the eaves, with the original doorway in the longer elevation facing south (towards the public road, not visible from the house), and an inserted door in the north elevation (Figure 23). There are no windows. The plain-tiled roof is pyramidal, with a modern louver.

### The brickwork:

The bricks are 8¾ x 4⅜ x 2½ inches, of pugged clay, hand-made and wood-fired. They are laid in lime mortar in Flemish bond on three sides, and English bond on the east side (facing away from the house and access road). The wall is 18 inches thick at the plinth, reducing to 13 inches seven courses above ground, with a weathered set-back on the outside and a plain step inside. There is a dentilled eaves course all round, similar to that described at 'South Suffolk B'. The mortar joints retain the incised lines which are characteristic of later eighteenth-century work. A neatly-cut inscription near the north-west corner may represent the initials of the builders:      J.P.   J.P.
                                                  P.-.      (last letter illegible)

### The doors:

The original oak door-frame is still present, including the threshold. The door is missing, but was 4 feet high by 2 feet 6 inches wide. Over it is a flat arch of *gauged* brick. A floor has been inserted 6 feet above ground, the lower storey becoming a cellar for storage. A doorway to the upper storey has been inserted in the north elevation; owing to the steep gradient of the site it stands only 3 feet above ground, approached by a short ramp.

### The former nest-boxes:

All the nest-boxes have been removed, but some evidence remains. In the upper room the walls are neatly plastered, but below the inserted floor the brickwork exhibits a regular pattern of holes in the vertical joints at every fifth course, indicating that battens of 2½ x 2 inch section projected to support the nest-boxes. One diagonal batten in a corner is still *in situ*, and many shorter stubs remain. If the material forming the nest-boxes was one inch thick they would have been 13

Figure 23: The dovecote of Polstead Ponds Farm, Polstead, from the south-south-east, with the original doorway, and the farmhouse beyond.

inches high and 15 inches wide. The 'shelves' would have been 11 inches wide, which may have included an alighting ledge. If the lowest tier was mounted on the 2½-inch set-back there were 9 tiers, with 38 boxes in each complete tier. Allowing for those omitted at the door there would have been about 338 in all.

### *The roof and pipe:*

A plaster ceiling has been inserted above the upper room, and subsequently removed. The original roof structure has survived intact, and what is more remarkable, the original pipe. Oak rafters of square section rise to a rectangular frame at the base of the former louver. Oak collars tenoned across two pairs of rafters enclose the pipe, which is made of pine boards, 1 foot 10 inches by 1 foot 7 inches in section, and 2 feet 6 inches deep. As with the others reported, the framing is formed outside to leave the inside smooth. Its function was described at Great Thurlow. This one differs from the others reported in having parallel sides.

### *Later developments:*

It is not clear what function the inserted upper storey served, but it is most likely that it became a granary. The floor structure is strong enough for that use, and the new doorway 3 feet above ground would have given as much protection against brown rats as many purpose-built granaries.

The Listed Building description of 1980 records that at that time the roof was slated. The present owner has re-clad it with hand-made red clay tiles (as it would have had originally), and has erected the present louver. It is square, with five *flight holes* of inverted-U shape in the south side, the other sides blank, and a felted pyramidal roof and weather-vane. Around its base is an inclined wooden ledge. This is an authentic detail which can be seen in many early illustrations. It provided a place on which the pigeons could perch when entering or leaving the building, and was inclined so that their droppings would be washed away by rain.

## KENTWELL HALL, LONG MELFORD  (TL 863 478)

Kentwell Hall is an Elizabethan manor house within a rectangular moat, set in a park. The eighteenth-century dovecote is outside the moat, 4 yards south of its south-west corner, and was formerly enclosed in a walled-in garden (Figure 24). It is 20 feet square and 18 feet high to the eaves. In the south elevation (facing away from the house, but towards the main approach road) is a central door with a window immediately over it, and three more windows just below the eaves. The plain-tiled roof is pyramidal, with a modern octagonal cupola. It is Listed as Grade II*.

This dovecote differs from nearly all the others described in having been conscientiously restored to usable condition - and is in fact used, although for keeping ornamental pigeons rather than for producing food. Owing to the extensive restoration it may not be possible to distinguish original features from later ones as positively as elsewhere.

### *The door and windows:*

The doorway is 6 feet 1 inch high by 3 feet 3 inches wide, rebated for a door opening inwards. The ledged door appears to be original, and is fitted with a stock-lock. The unglazed window over it is 1 foot 9 inches high and of the same width, with two mullions, below a shallow segmental arch. The three high windows are each 1 foot 4 inches high by 2 feet 10 inches wide, with two mullions. Holes inside the frames show that formerly they were protected against birds of prey by vertical bars 2 inches apart, but these original features have not been replaced. In 1936 Messent wrote: 'the side divisions . . . are glazed, but the middle ones are left open for ventilation'.[65] They are wholly glazed now.

Figure: 24: The dovecote of Kentwell Hall, Long Melford, from the south-south-east.

## The brickwork:

The bricks are 9 x 4¼ x 2¼ inches, of pugged clay, *kiln-fired*, and very regular in size. They are laid in lime mortar in Flemish bond; four courses rise 10¾ inches. The wall thickness reduces at two levels, with external 2-inch set-backs at 1 foot 8 inches and 4 feet 4 inches above ground. There is a projecting band of three courses 11 feet above ground. Most of the bricks are red, but some have been burned to blue. In parts of the side and rear walls this shows as a regular pattern of blue headers and red stretchers, but as it does not appear on the front elevation it may

Plate 5: At 'South Suffolk B', showing the nest-boxes of wattle and daub, the incomplete potence, and the beams supporting its upper bearing. The roof is modern.

Plate 6a (top): All that remains of the eighteenth-century dovecote at Crowfield Hall, Crowfield.
Plate 6b (bottom): The pigeon-tower at Scotland Place, Stoke-by-Nayland, from the west. The door to the pigeon-loft is in the far side, facing the River Box.

not be a deliberate embellishment. It is evidently intentional that the top course in each set-back and of the band is composed mainly of blue bricks.

**The nest-holes:**

There are 13 tiers of brick nest-holes, apparently integrated with the structural walls, with 44 to a complete tier (Plate 7, opposite page 72). Allowing for those omitted at the door and windows there are 526 in all. Each nest-hole is 11 inches high, 12¼ inches wide, and 13 inches from front to back. The entrance holes are 5½ inches high by 5¼ inches wide, all at the right side. They differ from most other dovecotes in being raised one course above the nest floors. Headers project to form a continuous alighting ledge to each tier, 3 inches wide. The lintel over the door is extended to form a continuation of the adjacent ledges. The lowest tier is 1½ feet above the brick floor. The construction is unusually substantial, comprising six courses of bricks to each tier, of which two courses form the floors. All the nest-holes are lined with clay daub, as was reported at Goswold Hall. All the brickwork has been lime-washed.

**The potence:**

The potence has been restored to usable condition. The original axis is a chamfered square oak shaft pivoted at the bottom in a wooden block, and at the top in an iron plate attached to the soffit of a transverse beam set in the highest course of bricks (Plate 7). The lower arm is original, 4¾ x 3½ inches, tenoned and pegged to the axis 4 feet above the floor, braced from above, with iron straps reinforcing the joints. The upper arm is of similar dimensions but restored, braced from below. The arms reach to within 2 inches of the alighting ledges; bolted to them is an old ladder with 23 rungs, steeply inclined. A practical demonstration showed that the ladder allowed easy access to every part of the building, including the nest-holes in the corners. A small platform has been built on the axis, level with the lower arm, on which a feeding box is now in use.

**The roof:**

The restored pyramidal roof is of shallower pitch than is usual with plain tiles. It has leaded hips and extended eaves which form integral guttering (and as no down-pipe is apparent, perhaps a convenient water supply for the pigeons). Short angle ties are fitted across the corners. Principal rafters rise from the corners and the middle of each side to a square frame forming the flight platform. Joggled purlins are tenoned and pegged to the principals, and common rafters are tenoned to them, though not all pegged (a form of construction comparable with that at Great Thurlow). A square trap in the flight platform has been fitted with a hinged trap-door, but there is no pipe. The ends of the rafters are cut to an ornamental profile and picked out with red paint. Parts of this structure are original. The present octagonal cupola has a domed roof, and differs entirely from the square turret shown in photographs published in an earlier guidebook of Kentwell Hall, and sketched by Messent in 1936. It retains the 'perching platform all round' which he noted.

*Additional information:*

Pevsner reported that this was 'said to be the largest in Suffolk'. Such claims need to be treated with caution - if only because he made the same claim for another dovecote at Saxmundham (which no longer exists).[66] The dovecote of Flixton Hall has 1,178 nest-holes, and that at Great Thurlow, although incomplete, retains evidence of nearly 1,000. Another incomplete dovecote at Eriswell, which will be described later, probably had about 1,800 nest-boxes.

A modern pit of full width now separates the doorway from the rest of the floor, and is crossed by a plank bridge. A small iron grating has been inserted in the front wall to right of the door arch.

About fifty white fantail doves were in residence when visited.

An estate map of 1613 marks a 'Dove House Meadow' on the east side of the house, but does not show a dovecote there or elsewhere.[67]

## SCOTLAND PLACE, SCOTLAND STREET, STOKE-BY-NAYLAND (TL 996 368)

The house is of sixteenth-century origin, ¾ mile north-east of the parish church. The pigeon-tower is 20 yards to the east-north-east, 3 yards from the River Box (which here is only a few feet wide), opposite a meadow. There is a pond 30 yards to the north-west. The farm buildings are west of the house, so the pigeon-tower is in a quiet and secluded position. It is of two storeys, 13 feet square, 16 feet high to the eaves, with a door facing north-west, and a ground-floor window to the south-east (Plate 6b, opposite page 69). Only the upper storey was designed for pigeons. The door to the loft is on the north-east side (away from the house). The plain-tiled roof is pyramidal, with a hipped dormer to the north-west and a modern louver.

*The brickwork:*

The bricks are 9½ x 4¼ x 2½ inches, of finely sieved and pugged clay, kiln-fired, mainly red, with some over-fired to blue. They are laid in lime mortar in English bond, meticulously executed so that the vertical joints align in alternate courses. Four courses rise 11¾ inches. Fine lines are incised in the mortar joints. A plinth stands seven courses above ground, terminating at a quadrant weathering 4¼ inches wide of rubbed bricks. Above the plinth the wall is 18 inches thick to first-floor level, reducing inside to 9 inches to form a step which supports the floor and nest-boxes.

*The pigeon-loft:*

The doorway to the loft is 9 feet above ground, accessible only by ladder. The ledged door is 6 feet 4 inches high by 2 feet 2 inches wide, opening inwards, secured from outside by a stock-lock, and from inside by a pivoted iron bar. The structure of the roof and dormer is original, of hardwood, wholly of nailed construction. Only the base of the original louver remains, with the skeleton of a former pipe; above the roof it is of modern construction. When first seen in 1991 this building was disused, and the louver had been closed with closely-spaced wooden slats. By 1998 it had been brought back into use for white doves, and the apertures had been re-opened and provided with inclined boards as shown.

## The nest-boxes:

The nest-boxes are built of clay bats, made of chalky yellow clay containing straw and small stones, of different thicknesses according to their positions (Figure 25). The side walls are 5 inches thick, the fronts are 3½ inches thick, and the floors are of bats 2½ inches thick which project to form alighting ledges 3½ inches wide to each tier. The boxes are 7½ inches high, 13 inches wide, and 10 inches from front to back. Each has an entrance hole of full height and 4½ inches wide. To left of the door, and on the south-east wall, the entrance holes are at the left side. All the other boxes which survive have entrance holes at the right side. Originally there were 9 tiers, with 25 in each tier, total 225, but those immediately to right of the door have been removed, so about 200 remain in good order.

Figure 25: Nest-boxes of clay bats at Scotland Place, Stoke-by-Nayland.

## The lower storey:

In 1991 the doorway to the lower storey was 6 feet 7 inches high by 3 feet wide, with an oak frame jointed with ¾-inch pegs and rebated for a door opening inwards (with renewed jambs), below a shallow segmental arch. By 1998 the frame had been removed. The window aperture to the rear is 3 feet 8 inches square, splayed at the jambs. It had been neatly blocked with early nineteenth-century brickwork, but has since been re-opened. The walls are plastered internally. The ceiling too was of plaster, 6 feet 10 inches high, but the woodwork was in poor condition, and has since been renewed, leaving the joists exposed.

Apparently the lower storey was originally designed as a garden pavilion for the ladies of the house. In the early nineteenth century it was converted into a dairy or cold larder, and it formerly retained a slatted bench for that purpose. It is now returned to its original use.

## Dating:
The use of English bond in this context is surprising, for Flemish bond was more common in fashionable buildings. Here the English bond is executed with unusual precision; it may have been adopted for its greater strength. Below the eaves are pine cornices moulded to an elegant classical profile. This is a late eighteenth-century building which combined domestic use as a garden pavilion with a loft for pigeons. Whether it was originally intended for ornamental birds, or for common pigeons kept for culinary use on a household scale is uncertain.

## Additional information:
A fourteenth-century dovecote in this manor is mentioned in Dr. Hoppitt's Appendix.

## THE PARK, GREAT GLEMHAM (TM 342 618)
The dovecote is at the west side of the park, near the original site of Great Glemham Hall. In 1814 the Hall was rebuilt 500 yards away on a high site in the middle of the park. The dovecote is isolated in a meadow, 40 yards north-east of the former stable block, near a tributary of the River Alde (Plate 8a, opposite page 73). It is octagonal, each face 7 feet wide, and 16 feet high to the eaves. The door faces south-east, with a round window above and a similar window opposite. The plain-tiled roof changes from a pyramid to a cone as it rises, with a modern turret or terminal.

## The brickwork:
The bricks are 9 x 4$^{3}/_{8}$ x 2½ inches, of finely sieved and pugged clay, uniform in size, laid in lime mortar in Flemish bond; straight lines are incised in the mortar joints. The plinth forms a weathered set-back 1½ feet above ground. The eaves cornice comprises three courses of bricks - one with a cavetto moulding, one plain course above, and one with a cyma recta moulding. On each side of the doorway vertical lines have been incised in some bricks, and filled with lime mortar to form a symmetrical pattern of false *closers*. The initials K H are inscribed in a brick on the right side of the doorway, 5 feet above ground.

## The door and windows:
The door-frame and door have been replaced in modern softwood to the original size, 6 feet 1 inch high by 2 feet 9 inches wide, leaving the jambs and flat arch of gauged bricks undisturbed. The threshold is of limestone. The two round windows are formed of gauged bricks. Each retains the original hardwood frame, and traces of two iron grills which have been installed at different times. The original grill was sunk in a rebate; nothing remains of it, but a series of holes at 2-inch intervals round the perimeter indicates how the bars were fitted. Over this is a round frame of wrought iron with the rusted remains of thin vertical rods one inch apart (Plate 8b, opposite page 73). In the north-east and south-west faces there are false windows formed with rings of gauged bricks in the same way, but enclosing only plastered brickwork. Traces of paint on the plaster show that it was painted to represent a black void within a white ring, so that from a distance the dovecote appeared to have four identical windows symmetrically arranged.

Plate 7: Inside the dovecote at Kentwell Hall, Long Melford, showing the restored potence and brick nest-holes.

Plate 8a (top): The dovecote in Great Glemham park from the south, showing the false window. The turret is modern. Plate 8b (bottom): The south-east window at Great Glemham, with remnants of a decayed wire grill.

## The nest-boxes:

Although the building is octagonal the nest-boxes are built round a cylindrical curve. There are 15 tiers of nest-boxes, with 36 in each complete tier, raised on a solid plinth which stands 1½ feet above ground. The pattern continues unbroken across the false windows. Allowing for those omitted at the real windows and doorway this makes 504 in all. Each tier is composed of four courses of bricks. Each nest-box is 9 inches high, 10 inches wide, and 9 inches from front to back. The entrance holes are 6½ inches high by 4¾ inches wide, all at the right side. Headers project to form an alighting ledge 2½ inches wide to each tier.

At the top, three courses of weathered bricks and intermediate tiles are laid to form a steeply inclined surface above the nest-boxes (Figure 26). This appears to be an ingenious way of denying any perching place to a bird of prey which might penetrate to the interior.

Figure 26: The brick nest-boxes at Great Glemham, built up with weathered bricks and tiles to avoid leaving any perching place for birds of prey. Drawing by Pamela McCann.

## The roof:

The roof structure is original, wholly of oak. Principal rafters of vertical section rise from the angles to a round frame at the top. Joggled purlins of horizontal section are tenoned and pegged to them, following the curvature. Common rafters of vertical section are nailed down to the circular wallplate, and bird-mouthed and nailed at the top. The tiles have been re-laid on modern laths. A simple modern turret is mounted on the round frame at the apex, designed only as a visual termination.

## Dating:

This is a building of high quality, designed by an architect and constructed by highly proficient bricklayers. The style remained in fashion throughout the Georgian period, but close examination of the brickwork indicates that it is of the early nineteenth century rather than the eighteenth. It may have been built soon after the Hall was rebuilt on another site in 1814.

*Later uses:*
   Inside, numerous small iron hooks have been driven into the mortar joints up to a height of 6 feet 3 inches. These have nothing to do with its main use as a dovecote, but relate to later use, probably as a game larder.
   The last use of the building was as a pump-house. Insulators for overhead cables have been left *in situ*, and a hole has been made for a pipe in the north-east wall. In the middle of the floor is a raised emplacement of hard yellow bricks for a pumping engine, with a half-floor to give access to it. A disused water tank remains in the roof, which must have been built *in situ*, for it is much too large to pass through the door. If a potence was formerly present this installation has destroyed any remaining evidence.

*Additional information:*
   The present Hall was built for the Reverend Samuel Kilderbee, Rector of Campsea Ash, who inherited the estate from his father, Samuel Kilderbee, an Ipswich lawyer who was a friend of Thomas Gainsborough, in 1813. He sold the estate in 1829.[68]

## FOUR BRIDGES, LETHERINGHAM (TM 282 585)

The dovecote is situated on the southern edge of a meadow through which the River Deben flows, part of Easton Park, and 30 yards north-east of the house called Four Bridges. It is octagonal, and originally it differed in only minor respects from the dovecote in Great Glemham park. It appears very different now because it has been altered to another use since it ceased to fulfil its original function. Short spur walls have been built out to the south to enclose double vehicle doors (Figure 27). It has similar round windows to north and south, but no false windows.
   Otherwise the differences are minimal, so it need not be described in full. A green deposit makes the brickwork appear different, but where this has been rubbed away by cattle the bricks are red. The gauged brickwork round the windows is, if anything, better executed than at Great Glemham. The north window has been blocked inside, but the iron grill is identical with those at Great Glemham, and the degree of rust decay is similar too. A modern grill has been fitted in the south window. Frogged bricks are visible where they bridge the entrance holes of the nest-boxes. There is no inclined brickwork round the wall-head. A floor has been inserted at mid-height. The roof has been rebuilt with modern softwood; there is no cupola, but there is a weather-vane. Some repairs inside have been executed with flettons and cement mortar. The ground floor is of concrete.
   The bricks which form the spur walls are the same as those of the main fabric, and they are laid in lime mortar, so although it is now used as a garage it was converted to vehicle storage in the nineteenth century.

*Dating:*
   Over the south window is an inscription: 18 R D 2 - (last digit illegible) The associated house, Four Bridges, is dated 1829, so probably the final digit was a 9.
   This is only 4½ miles south-west of the dovecote at Great Glemham. Apparently the two buildings were built to the same design, but were executed slightly

Figure 27: The dovecote at Four Bridges, Letheringham, from the south, now used as a garage. Easton Park is beyond.

differently, perhaps by different contractors.

In a wheat-growing area 1829 would be a late date for the construction of a dovecote, particularly one with so many nest-holes, for by then the traditional practice of allowing large numbers of pigeons to find their own food was reckoned to be uneconomic. However, as the Raynbirds described, 'the agricultural revolution' developed late in Suffolk, and at that time it was still largely pastoral.

*Additional information:*
   Messent described this building in 1936, wrongly locating it in the parish of Easton. It was much the same as at present, except that then it had 'a cylindrical cupola and finial'. The walls were covered by ivy, the roots of which had eaten deeply into the mortar joints.[69]

## HITCHAM HOUSE, HITCHAM  (TL 978 507)
This house was built as the rectory of Hitcham. It is situated half a mile south-west of the parish church, facing towards it. The pigeon-tower is 65 yards south-west of the house and slightly uphill from it. It is 12 feet square, 11½ feet high to the eaves, with a door to the south-east (obliquely visible from the house), high windows to front and rear, and an additional low window to the rear (Figure 28). The plain-tiled roof is pyramidal, with a square louver. It was designed as a two-storey building from the outset, with only the upper storey allocated to pigeons. It is not Listed.

*The brickwork:*
   The *gault* bricks are of fine quality, 9 inches long by 4½ inches wide by 2½ inches high, laid in lime mortar in Flemish bond; four courses rise 11½ inches. The eaves cornice comprises one course of moulded bricks inclined outwards at 45 degrees, one course of separate headers forming dentils, and one course aligned with their ends.

*The door and windows:*
   The doorway is 5 feet 11 inches high by 2 feet 8 inches wide, and has an original ledged door with strap hinges, opening outwards. Over it is a course of headers on edge supported on the lintel. The two high windows are immediately below the cornice, 1 foot 5 inches square. When first inspected in 1992 they had original wooden frames, and were protected against predators by cast iron grills with 15 vertical slots half an inch wide. Since then they have been replaced with modern materials.

*The roof and louver:*
   Short angle-ties are halved across the wallplates. Above wallplate level the roof structure is wholly of softwood, of nailed construction. There are seven parallel rafters in each side, in addition to the hip rafters, rising to a square frame which supports the louver. The roof is clad with red plain clay tiles, with leaded hips. In 1992 the louver had two tiers of eight flight holes of semi-circular shape, with a perching ledge on ornamental brackets between them, and a leaded pyramidal roof with a plain wooden finial.

*The interior:*
   There is an original floor 6 feet above ground, supported on a main beam of 7 x 6 inch vertical section resting on wooden pads sunk in the side walls. There must have been a trap-door to the pigeon-loft, but in 1992 this was inaccessible. All the nest-boxes had been removed, leaving only stains on the brick walls. There seem to have been just over 200 nest-boxes 10 inches wide, but little else could be deduced.

Figure 28: The pigeon-tower of Hitcham House, Hitcham, from the south-south-east.

The original purpose of the lower storey is unknown, but probably it was a tool-shed or miscellaneous store.

## *Dating:*

In May 1814 James Spiller of Guilford Street, Middlesex, drew up a report on the condition of Hitcham Rectory (which at that time was a timber-framed building), and a proposal for rebuilding the walls in brick. The work was to be carried out for the Reverend John Staverton Mathews. James Spiller was a pupil of James Wyatt, and later worked occasionally with Sir John Soane.[70] The present house, which appears to be the outcome of that proposal, is of two storeys, with two full-height bows, curved sash windows, giant pilasters of brick, and a small pediment. The brickwork is the same as that of the pigeon-tower; probably both were built by the same architect at the same time.

### Additional information:
In 1992 this pigeon-tower was neglected and in poor condition. By 1996 it had been restored to use, with about a dozen white fantails in residence.

## COLDHAM HALL, STANNINGFIELD (TL 864 556)
Coldham Hall is a major brick house dating from 1574. The dovecote is isolated 140 yards to the south. There is a lake 130 yards to the north-west of it. The dovecote is 23 feet in diameter, 26 feet high to the eaves, designed in an ornate Italianate style (Plate 9a, opposite page 88). The door is to the north-north-west, enclosed in an elaborate porch; the only window is to the south-south-east. The plain-tiled roof is conical.

### The exterior:
A plinth rises six courses vertically; above this the brickwork *batters* to a height of 9 feet 3 inches, where there is a projecting band comprising two courses of gault bricks and a weathered course of red bricks. Above this the wall rises vertically. Five false windows and one real window, all with arched heads and outlined with gault bricks, are symmetrically disposed round the building. The false windows are plastered inside. Above them the upper part of the wall projects on a band of corbelled round arches made of gault brick which resemble machicolations, with a ball pendant of limestone below each corbel. At the rear three arches are left open to serve as flight holes (Figure 29). Above this is a wide coving of lath and plaster to the eaves.

### The brickwork:
Most of the structure is formed of red bricks 9 - 9¼ inches long by 4½ inches by 2½ inches, of finely sieved and pugged clay, laid in lime mortar in Flemish bond; four courses rise 12 inches. In the conical curvature of the lower part it is inevitable that the headers become slightly out of line in places, but some flexibility was achieved by using oversize bricks. The ornamental details are executed in gault bricks of fine quality, 9 x 4½ x 2½ inches, and some limestone.

### The door and window:
The heavy oak door is 6 feet 1 inch by 2 feet 4 inches, with a two-centred head and ovolo-moulded fillets. It opens inwards, and has an elaborate set of bolts of the full width of the door, secured by screwed spindles, with keyholes in the middle for locks which are now missing. It is set deeply inside a gabled porch of red brick and limestone with an arch of similar shape. The threshold is of limestone. One jamb has been reinforced by a massive oak post, bound to it with iron straps.

The arched window is 3½ feet by 1½ feet, protected by a wrought iron grill formed of vertical strips 1 inch wide and 1¼ inches apart (Figure 29).

Figure 29: The barred window and arched flight holes at Coldham Hall, Stanningfield. Note the timber projecting through the wall at right, and four holes made in the brickwork at the same level to support the floor of the inserted loft.

## *The nest-boxes:*

The inside is cylindrical. The nest-boxes are mounted on a brick plinth 1½ feet above ground, and are formed of L-shaped blocks of chalky clay containing straw, 3¼ inches thick, assembled with lime mortar (Figure 30). The floors are of clay bats 2½ inches thick, which project to form an alighting ledge 2½ inches wide to each tier. These floor bats are finely profiled to follow the curvature of the building. 22 tiers of nest-boxes are *in situ*, with 37 boxes in each complete tier, so allowing for those omitted at the door and window there are 774 nest-boxes. (Above the highest tier there is sufficient unused height for another three tiers, so there may have been 885 nest-boxes originally). Over the door and window the nest-boxes are carried on curved hardwood lintels which follow the line of the adjacent alighting ledges. The inner surface of the brick wall, which forms the backs of the boxes, is plastered. The whole assembly is a highly accomplished piece of design and workmanship - perhaps the finest construction in clay bats in Britain (Plate 9b, opposite page 88).

Figure 30: Nest-boxes of clay bats at Coldham Hall, Stanningfield, shaped to follow the curvature of the wall.

### *The potence:*

There is an incomplete potence, comprising an axis of octagonal section mounted on a similar wooden post which supports the lower bearing 2 feet above ground. The upper bearing is in a beam across the top of the brickwork, braced by two spur beams to each side. (Plate 9b.) Only the upper arm is still present, bracketed from above and below.

### *The roof and cupola:*

The structure of the roof is concealed by lath and plaster to the soffit. When inspected it had been damaged by a storm; there was only a hole where the cupola had been. Messent described it in 1936: 'The top is crowned with a cupola which is supported on six legs and has a circular cornice and hemispherical dome with knob pinnacle'.[71] Salvaged components stored within the building included a wooden frame 4 feet in diameter, and the six legs. It is understood that the storm damage has been repaired since. Whether the cupola ever served any purpose other than ornament is not clear, for the present pigeon entrances are the three arched apertures in the corbelled band.

## The entrance loft and pipe:

Inside these arched flight holes is a decayed wooden loft about 5 x 5 x 5 feet. It is mounted on beams supported in the brickwork, and hung from the spur beams above. It was built after the original construction, for holes were broken through the clay nest-boxes and the outer brickwork to insert the beams; one beam projects outside (Figure 29). From the floor of the loft a wooden pipe descends, 2 feet 5 inches deep, flaring slightly to form a lower aperture 2 feet square (Plate 9b). Apparently the original entrance for the pigeons was found to be unsatisfactory - perhaps too vulnerable to birds of prey - and the loft and pipe were installed to provide better protection.

## Dating:

The Listed Building description of 1984 dates this dovecote as 'mid-nineteenth century, with an earlier core'. Certainly the lower stage appears to be different from the rest, but that is because the inclined bricks are more exposed to drip from the eaves, and so have acquired a rich growth of lichen. Close examination shows that the brickwork there is exactly the same as that above. It is unlikely that this dovecote was built so late, at a time when large-scale pigeon-keeping was passing out of use. This is confirmed by the tithe map of 1838, which records a round building marked 'Dove House' in exactly the same position. Therefore it was built early in the nineteenth century, at some date before 1838. The field boundaries have changed since, for then it was in the middle of a 14-acre field named 'Dove House Meadow', while now it is near the corner of an orchard.

The unusually stout door and its elaborate locking arrangements must have been installed for defence against the large-scale thefts of live pigeons which were common at the time. The strangely massive repair to one door-post suggests that there was at least one attempted break-in. The window too is more strongly barred than is common elsewhere. It is not clear why this dovecote was built so far from the house, where it was exposed to theft, but ostentatious display is certainly a possibility.

## Additional information:

Coldham Hall was occupied by the Rookwood Gage family, baronets, throughout the eighteenth century. Robert Rookwood Gage was still there in 1819, succeeded by his brother John Rookwood Gage (1786 - 1842), who let the property in the 1820s and 1830s.[72] The tithe apportionment of 1838 records that the occupier was James Talbot, who farmed 240 acres in the parish.

## COCKFIELD HALL, YOXFORD (TM 396 692)

Cockfield Hall is a major house which externally is in early nineteenth-century Tudor style, although it is of much earlier origin. The dovecote is in the middle of an outer court formed by brick buildings of various periods, all either Tudor or in Tudor style - a chapel, a barn, a dairy range, and a gatehouse leading to an inner court and the Hall itself. The dovecote is similar in style, octagonal with angle pilasters at 8 feet 9 inch centres, 21 feet high to the parapet (Figure 31). The door

Figure 31: The dovecote of Cockfield Hall, Yoxford, from the south.
Photograph by courtesy of Tim Buxbaum.

is to the north, and there are windows to east and west. The roof is pyramidal, with an octagonal cupola.

### The brickwork:

The bricks are 9 x 4¼ x 2½ inches, of high quality and very uniform in size, laid in lime mortar. The integration of the nest-holes with the outer fabric produces a curious pattern consisting of Flemish bond interrupted at intervals by irregular courses. In the lower two-thirds there are four courses of Flemish bond to one course of headers, and in the upper third there are three courses of Flemish bond to one course of headers. From ground level the change of pattern is barely perceptible.

### *The door and windows:*
The chamfered door-frame is of hardwood, below a segmental arch formed of plain headers with tapered mortar joints. The ledged and boarded door is of softwood, of normal domestic size, and may be original too. The unglazed windows are 7 feet above ground, 3 feet high by 3½ feet wide, with similar arches. They are protected from predators by wooden slats 1 inch wide and 1½ inches apart, arranged in chevron pattern.

### *The nest-holes:*
The interior is octagonal (Figure 32). The nest-holes are above a solid plinth standing 2 feet 8 inches above ground. The first nine tiers are composed of five courses of bricks, forming holes 9¼ inches high by 9 inches wide by 10 inches from front to back, with an alighting ledge 3 inches wide to each tier. The entrances are 7 inches high by 4¼ inches wide, all offset to the left. There are 44 nest-holes in each of these tiers. Above them are four courses of nest-holes composed of only four courses of bricks. In these tiers an extra nest-hole is fitted into each corner, so there are 48 in each tier. These higher nest-holes are of the same size and pattern, but are differently constructed in that their floors are composed of only one course of bricks, instead of two courses as in the lower tiers.

In the middle is a cylindrical plinth 4½ feet high which supports the lower bearing of a potence. This accommodates two more tiers of 24 nest-holes, each formed of four courses of bricks; they are 11 inches wide at the front, reducing axially towards the rear, with entrance holes 7 x 4¼ inches. Allowing for those omitted at the door and window this makes a total of 608 nest-holes, all still in good order.

### *The potence:*
The potence is of oak, original and in usable condition - a rare survival (Figure 32). The axis is square at the top and bottom, chamfered with run-out stops to an octagonal section. Two bracketed arms emerge from adjacent faces, tenoned and pegged to support an inclined ladder; the lower arm is additionally supported by an iron stay. The upper bearing is in a wooden pad at the crossing of two chamfered tie-beams.

### *The parapet, roof and cupola:*
The parapet is crenellated, with flat stone copings. Above each pilaster is a tall pinnacle of moulded brick. The pyramidal roof is of the normal pitch for tiles, simply constructed of softwood rafters with purlins at mid-height, all nailed. Between the roof and the parapet a leaded gully is formed of boards, draining into two cast iron down-pipes.

The cupola appears to be original. It has eight hardwood posts of octagonal section supporting an octagonal frame, with a plain-tiled pyramidal roof and a

Figure 32: Inside the dovecote of Cockfield Hall, Yoxford. The potence is original and in working order. The top four tiers of nest-holes are shallower than those below.

spiked wooden finial. There are no other protective devices against birds of prey. In a busy courtyard in an estate staffed by gamekeepers who trapped and poisoned every kind of predator there may have been little need for extra protection against birds of prey.

About a dozen white fantail doves were in residence when visited.

### *Dating:*
The Listed Building description dates this as mid-nineteenth century, but that seems improbably late for a new dovecote accommodating so many pigeons. It may be more reliably dated by an associated building. The main access road to Cockfield Hall leads from the middle of Yoxford, immediately opposite the parish church. The north aisle and north chapel were built in the same Tudor style in 1837. At that time the estate was owned by Sir Charles Blois, who inherited the baronetcy and the estate in 1810, and who died in 1850. The architectural parallels are so close that it is likely that the dovecote was designed by the same architect at about the same time.

# Chapter 7
# Stone dovecotes

**THE DOVECOTE, 5 CHURCH WALK, MILDENHALL** (TL 709 745)
This former dovecote is now a roofless ruin, scheduled as an Ancient Monument. It is situated in the garden of a private house, 200 yards west-south-west of the parish church. The River Lark is 250 yards to the south. The remains form a rectangle 39 feet by 26 feet externally, orientated north-south (Figure 33). The original doorway is in the middle of the east wall, and there is a modern doorway in the north wall. The building has been roofless a long time and is much eroded on the inside, but in places the wall still stands 10 feet high. No evidence of windows remains.

*The fabric:*
The main fabric is of flint and clunch rubble, forming walls which originally were 3 - 3½ feet thick, with thin courses of grey sandstone at 6-inch vertical intervals. The quoins are of Barnack stone, varying in depth from 9 to 12 inches, irregular where they abut on the rubble, but so finely dressed elsewhere that the horizontal joints bed perfectly with little or no mortar. The outside was rendered originally, but only traces of the render remain.

*The doorways:*
The early doorway in the middle of the east long wall is 5½ feet high by 3 feet wide, with a four-centred arch; the jambs are slightly splayed towards the inside, with some modern repairs. The north doorway is in similar style but is wholly modern.

*The nest-holes:*
The nest-holes were formed between the layers of sandstone, which supported the rubble above, so they were 6 inches high throughout. In plan they were club-shaped, as was common in medieval dovecotes; that is, they were 5 - 6 inches wide at the entrance, opening out inside to form a rounded chamber 10 - 12 inches wide. The inner faces of the walls are so eroded that only the remoter parts of the nest-holes remain. Nevertheless, they exhibit the characteristic shape, and some retain an inner surface of lime mortar. The lowest tier of nest-holes is one foot above present ground level, but this has risen substantially, so it may have been two feet or so above ground when built. Twelve tiers of nest-holes are still visible. They are so eroded that an exact count is impossible, but measurements indicate that there were 72 - 75 nest-holes in a complete tier. Even with walls of the present height that makes 850 - 900 nest-holes. It is not known how high the walls stood originally; in 1926 they were reported to be 12 feet high in places.[73] By comparison with other medieval dovecotes it is likely that they were high enough to accommodate 1,100 to 1,300 nest-holes.

*The historical background:*

The Ancient Monument schedule says of this dovecote 'perhaps once belonging to Mildenhall Priory', but there was no priory at Mildenhall. It is almost certainly in the grounds of the former hall. The Abbot of St. Edmundsbury owned the manor of Mildenhall, so this was his dovecote. Claude Morley proposed that it was built 'about 1290; it is certainly before 1400'. This dating would make it the earliest remaining dovecote in Suffolk. Since the Dissolution the property has been in secular hands - successively the North, Hammer and Bunbury families.[74] So little information survives that one can only place this building in the context of monastic property in general. That is, when first built it would have supplied squabs directly to the Abbot and his entourage. The diet which wealthy potentates and their households enjoyed was described in the detailed accounts of the Bishop of Hereford for 1289-90 (although he was less wealthy than this Abbot). The year was divided into meat days and fish days. On meat days two or three kinds of fresh meat were served, summer and winter alike, supplemented by two or three kinds of poultry or wildfowl. During Lent and on fish days several kinds of fish and 'white meat' (cheese) were substituted. Pigeons were served on meat days from Easter to early November. Wines and imported foods from all over the known world - dried fruit, citrus fruits,

Figure 33: The remains of the medieval dovecote at Mildenhall, facing south. The nest-holes were formed in the white clunch, between the layers of grey sandstone. Erosion has destroyed many of them, but some can still be seen in the corners.

sugar and spices - were served as a matter of course.[75] For the wealthy there was no shortage of any food at any season.

During the fourteenth century great lords and institutions leased off their manors, and used the revenue to buy the produce which had formerly been supplied by the dependent granges. This change had begun before the Black Death of 1349, but it became general practice afterwards. From that time this dovecote would have been operated by the lessee, and much of its produce went to market. After the Dissolution in 1539 the dovecote may have continued in use in the same way, but at some stage - then or later - it was probably converted to other purposes. Often a later use can be deduced from the size of the doorway and other secondary features, but in this case there was already a breach in the north wall before it was taken into national guardianship, so nothing is ascertainable.

## THE ABBEY DOVECOTE, THE ABBEY GARDENS, BURY ST. EDMUNDS
(TL 856 642)

In 1920 Arthur O. Cooke mentioned 'the so-called "dovecote", the remains of which will be seen among the abbey ruins at Bury St. Edmunds', and said: 'Of any sign that it was formerly a dovecote there is none'. This doubt was repeated by M. J. A. Beacham as recently as 1990.[76] The evidence is so easily visible that it is difficult to see how they came to that conclusion; apparently neither of them went there. The building is Scheduled as an Ancient Monument and Listed as Grade I, and in both cases is described as the Abbey Dovecote.

### *The site:*

What remains is a roofless hexagonal tower and a fragment of wall, now isolated in the Abbey Gardens, 17 yards west of the River Lark (Figure 34). A plan displayed by English Heritage nearby shows that originally it was one of two similar towers which formed the north-east and south-east angles of the Abbot's garden; nothing survives above ground of the other (Figure 35). A short length of rubble wall west of the tower is the only part remaining of the wall which formerly enclosed the garden. The leat to the Abbey mill has been filled in, but still shows as a shallow depression in the ground; it was immediately outside the east wall of the garden, turning westwards round the surviving angle-tower. The mill was 40 yards to the north-west.

### *The fabric:*

In plan the tower forms a slightly irregular hexagon, a little over 9 feet across each face. The upper part has fallen, but even as a ruin it still stands over 21 feet high. The fabric is of flint rubble with quoins and a weathered plinth of Barnack stone. In the south side there is a narrow doorway. In the west, north, and north-east sides there are original plain *loops* 6½ feet above ground facing outwards, formed of Barnack stone and deeply splayed inside. The south-west wall has in the upper storey a window inserted in the fifteenth century, with a segmental arch and moulded label. Above the doorway there is a mullioned and transomed window inserted in the sixteenth century. As with all monastic property much has happened to this building since its first use, but in this publication we are concerned primarily with what remains of the earliest phase, as a pigeon-tower in the boundary wall of the Abbot's garden.

### *The interior and the nest-holes:*

Inside the walls there is a set-back of 1½ feet about 11 feet above ground, dressed with stone slabs, which formerly supported a floor. The rectangular loops are rebated on the inside for shutters or other protective devices. The thick walls leave so little space in the lower storey that it cannot have been of much use except for defence and for storage. The evidence that the upper storey was intended for use as a dovecote is clearly visible from outside the barrier which now blocks the doorway. The remains of rectangular nest-holes 9 inches high by 12 inches wide are outlined by red bricks 2 inches thick, incorporated in the rubble fabric (Plate 10a, opposite page 89). Between the tiers of nest-holes are courses of red tiles which

Figure 34: The pigeon-tower which was formerly in the wall of the Abbot's garden of St. Edmundsbury Abbey, and a fragment of wall to the right, from the north.

Plate 9a (top): The dovecote of Coldham Hall, Stanningfield, from the north-west.
Plate 9b (bottom): At Coldham Hall, showing the incomplete potence, the loft inserted inside the three flight holes, and the pipe descending from it.

Plate 10a (top): At the Abbot's pigeon-tower, St. Edmundsbury Abbey: remains of brick nest-holes embedded in the fabric of flint rubble.
Plate 10b (bottom): The combined stable and pigeon-loft at Home Farm, Drinkstone, from the south-south-east, with the farmhouse beyond.

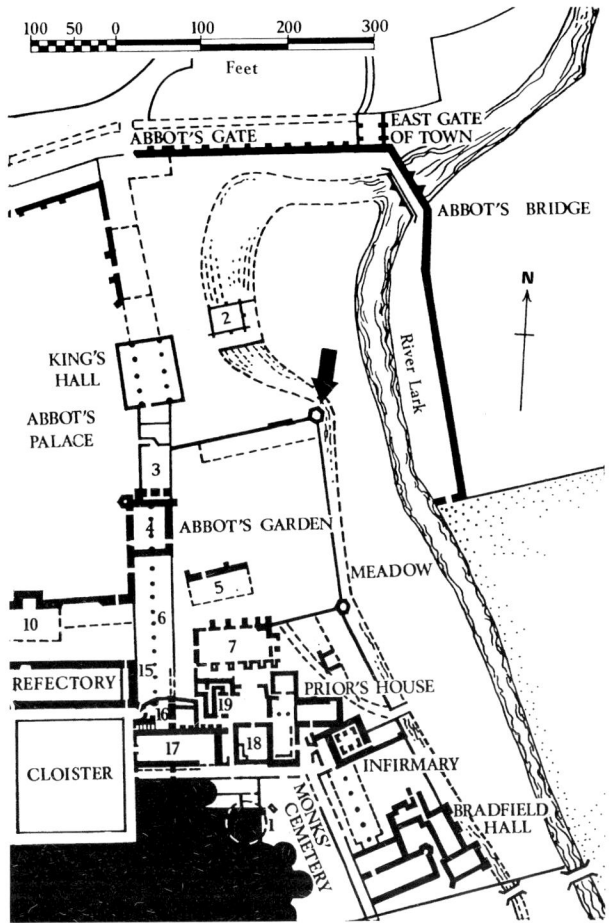

Figure 35: Provisional plan of part of St. Edmundsbury Abbey. The existing pigeon-tower is marked by an arrow; the mill is numbered 2.
Based on a plan by A. B. Whittingham, by courtesy of English Heritage.

served to support the rubble while the slow-setting lime mortar was hardening. All the Abbey buildings have been extensively robbed of usable building materials. This tower has been less stripped than some, but every accessible brick has been taken away, leaving only those which are so deeply embedded in the rubble fabric as to be not worth removing. The remains of the nest-holes are clearest in the three walls opposite the doorway. They are less clear in the other walls; two walls have been extensively altered for windows which were inserted later when the tower was converted to other use. Even so, the courses of tiles are clearly visible on the outside of the south-west wall beside and above the inserted window. Similar evidence is visible in the north-west wall, though less conspicuous there.

### *Dating:*

English Heritage has assigned this building to the fourteenth century. At that period brick was available, but it was still an expensive product, used only in prestigious buildings. It is worth noting that the angle-towers were fortified, and projected to provide clear sight lines outside the walls. Relations between the Abbey and the townsfolk were often turbulent, and there were major insurrections against the power of the Abbot in 1327 and 1381. He had good reason to form his garden as a defensible enclosure within the outer perimeter wall.

### *Later use:*

The inserted windows already described destroyed many of the nest-holes, and converted this building into a garden pavilion or gazebo. Probably the others were blocked then. The English Heritage guidebook reproduces two engravings which illustrate this building in the eighteenth century. One made from a painting of 1725 shows it prominently, the other dated 1741 more distantly, but in both it is depicted with a pyramidal roof and hexagonal lantern.[77]

## MARSH STABLES, CHURCH STREET, EXNING (NEWMARKET) (TL 621 654)

Exning is in the Western Sand region, only just in Suffolk - at the western edge of a small protrusion into Cambridgeshire. In 1804 Arthur Young wrote: 'Pigeon houses abound in the open field part of the county bordering on Cambridgeshire'.[78] If they abounded then, nearly all of them have disappeared. It is 100 yards south-south-east of St. Martin's Church; the trunk road A45 (built 1973) passes immediately to the south-east on a high embankment. The surroundings have changed so much that it is difficult to reconstruct the former context. The building is 19 feet square and 13 feet high to the eaves, with an original doorway (now blocked) and a modern window to the west-north-west, a modern doorway opposite, and a modern window to the south (Figure 36). The plain-tiled roof is hipped, with large gablets to north and south. Exning is in horse-racing country, on the edge of Newmarket; the building is now used as a loose box for a race-horse.

### *The exterior:*

The main fabric is of *clunch* in quarry-faced blocks up to 14 x 6 inches at the exposed face, with larger blocks as quoins. Clunch is not a durable material when exposed to the weather; here the surface has receded an inch or two by erosion, which suggests that originally it was protected by rendering. Bricks are used for the plinth, the eaves cornice, and the dressings of the west doorway. They are 9 to 9½ inches by 4¼ inches by 2¼ inches, of unsieved and unpugged clay, irregular in shape and burned to various colours - red, plum and brown - and laid in lime mortar in irregular bond, four courses rising 11¼ inches. The simple eaves cornice comprises one *dog-tooth* course supporting one plain course. The top of the wall has been tapered off with tiles below the eaves, but this may not be original.

### *The door and windows:*

The original doorway was 3 feet 2 inches wide, with a window immediately above of the same width. The frame has gone and the aperture has been blocked, leaving only a modern window. The brick dressings show that together they were 7 feet 4 inches high. The inserted stable door in the east wall is 6½ feet high by 4 feet wide.

Figure 36: The dovecote at Marsh Stables, Exning, from the north-west.

## *The nest-boxes:*

The north and south walls have been much altered, but the other walls retain visible evidence of the former nest-boxes, all of which are now blocked. They were formed in an inner structure of brick, with four courses to each tier. The entrance holes were 6 inches high by 5 inches wide, 10½ inches apart. There was an alighting ledge of headers to each tier; all of them have been roughly hacked back to the adjacent surface. Each box was 8½ inches high by 11 inches wide. The total thickness of the clunch wall and the brick nest-boxes is 2 feet 3 inches. Eight tiers of blocked nest-boxes are visible; a dado of thick plaster to a height of 5 feet conceals the lower tiers, but probably there were 12 tiers, with 44 - 48 boxes in each complete tier. Allowing for those omitted at the door and window there would have been 500 - 550 nest-boxes originally.

*The roof:*

The roof is simply constructed of hardwood rafters and collars of 4 x 3 inch vertical section. There are old lath-nails in the soffits at six-inch intervals. Above the collars is a wooden ceiling in the same position as the former flight platform, but it is of machine-sawn softwood. The gablets which formed the pigeon entrances are boarded over. The roof is clad with plain tiles of many tones and colours - cream, buff, pink, and brown, more characteristic of Cambridgeshire than of Suffolk. The Listed Building description assigns this building to the seventeenth or eighteenth centuries, and one can see why it is so unspecific. Apart from the eaves cornice it is completely plain, exhibiting no datable features. This type of roof was used on dovecotes for over four centuries. The bricks seem more characteristic of the seventeenth century than the eighteenth, but at all periods there were inferior and superior qualities of bricks, and sometimes old bricks were re-used in a later building.

## OUSDEN HALL, OUSDEN (TL 736 597)

Ousden Hall, a large country house near the parish church, was demolished in 1955, leaving only the service range and a free-standing clock-tower. The dovecote is in a paddock on the opposite side of the road from the service range (Figure 37). There is a lake 200 yards to the south. The building is 17 feet square and 12 feet high to the eaves, with a door to the west (facing the road and service range) and a round window to the south. The slated roof is pyramidal, with a square louver. It is now used as a stable.

*The walls:*

The fabric is of uncoursed large flints and lime mortar, forming walls 12 inches thick. The flints are worked only sufficiently to prevent the largest ones from protruding beyond the vertical plane. The dressings and toothed quoins are of pugged and kiln-fired red bricks of high quality, 8¾ x 4⅛ x 2½ inches, laid in lime mortar; four courses rise 12 inches.

*The door, window and decorative roundels:*

The doorway is 6 feet high by 3 feet 3 inches wide; the aperture is unaltered but the frame has been renewed in machine-sawn softwood, with a modern stable door. The threshold is of an early form of concrete, much like stone in appearance. Above the door is a roundel of brick containing a re-used limestone panel on which a lion passant is carved in high relief. The south window is 7½ feet above ground, 2½ feet in diameter, now blocked with flettons. In the north and east sides there are false windows formed with similar roundels of brick; the blank spaces inside are set back 2 inches and cement-rendered.

*The interior:*

The inside is smoothly plastered to a height of 6 feet, and above that the surface is concealed by many coats of lime-wash. The nest-boxes have been removed, but at irregular intervals in the fabric one can still see wooden plugs 2 x 2 inches, secured in place by wedges. Apparently the structure of nest-boxes was built up

Figure 37: The dovecote of Ousden Hall, Ousden, from the south-west.

from the ground, not dependent on these plugs except for stability. Nothing more can be deduced about them, but the wall height would be sufficient for ten or eleven tiers of nest-boxes one foot high. The interior is 15 feet square, so if the nest-boxes were one foot wide there would have been 58 in a complete tier. Allowing for those omitted at the door and window this tentative calculation suggests a total of 580 - 638 nest-boxes. The present floor is of small yellow bricks on edge, probably laid when the building was converted to a stable.

### *The roof, louver and former potence:*

The roof is simply constructed of softwood rafters with joggled purlins. At the top the rafters are bird-mouthed to a frame 1 foot 10 inches square which forms the base of the louver. The roof is now clad with Welsh slates, but it is of the steeper pitch which was designed for plain tiles. The structure of the louver seems to be largely intact, but the sides have been boarded over; its roof is of lead. There is no evidence of a pipe. Two beams of hand-sawn oak mounted on the wallplates cross in the middle; a hole of one-inch diameter at the crossing is the only evidence remaining of a former potence. (It was determined at Kentwell Hall that a potence is practicable in a square building). The style, and the quality of the brickwork, suggest that this dovecote was built in the early nineteenth century.

*Additional information:*
The free-standing clock-tower already mentioned is about 50 yards to the south-west. It has been occupied by feral pigeons, although whether it was designed as a pigeon-tower could not be determined. It is square, of red brick, built in three stages. In the third stage there are recessed panels; within them is an emplacement for a clock (now missing) on one side, and round windows on the other three sides. At mid-height there is a projecting band round three sides. The only access to the upper part is by ladder to a door in the middle stage. The leaded pyramidal roof is inside a parapet, rising to a square turret with tall arched apertures on each side, and a weather-vane. This tower is much later in date than the flint dovecote, typically Victorian. It is so much higher that it would have been more attractive to pigeons than the relatively squat dovecote described.

## HOME FARM, DRINKSTONE  (TL 958 617)
Home Farm is 150 yards north-west of the parish church. The farmhouse faces south-east into the farmyard; it is of the seventeenth century, with later alterations. The pigeon-tower stands 25 yards east of it (Plate 10b, opposite page 89). It is 15 feet square and 14 feet high to the eaves, in two storeys, with doors to both storeys facing south-west. The plain-tiled roof is hipped, with gablets to front and back. The farm pond is 8 yards to the north-east.

*The fabric, doors and window:*
This is a simple functional building, built from the outset as a stable, with a pigeon-loft above. The walls are of roughly coursed rubble - undressed flints and field stones, including some bricks - with brick quoins. An aperture of full height 4 feet 2 inches wide with brick dressings accommodates a modern stable door, a blocked window above, and an old door to the loft. There are no other apertures. The bricks vary slightly in size, about 9 x 4¼ x 2½ inches, of pugged clay containing small stones, laid in lime mortar. A lean-to extension of similar materials has been added to the north-west.

*The pigeon-loft:*
The floor is 10 feet above ground, with a trap for access in the south corner. The door to the front is 3 feet 4 inches high by 3 feet 10 inches wide. The masonry walls are plastered internally, and have been lime-washed.

*The nest-boxes:*
Most of the nest-boxes have been removed; only seven remain. They are made of ¾-inch pine boards, 9½ inches high, 11 inches wide, and 10 inches from front to back; each has an entrance hole of inverted-U shape 7 x 4 inches close to the left side. There is sufficient space on the walls for four tiers of up to 42 boxes, a maximum of 168 nest-boxes of this size.

*The roof:*
The wallplates and roof are made entirely of re-used oak from a medieval building, in clasped purlin form with rafters of horizontal section (as they were in their first use), secured by nails. Two pairs of long rafters form the gablets. There is no evidence of a louver, flight platform or pipe. The front gablet has been boarded over, and now has three flight holes of inverted-U shape. The rear gablet is closed with laths and cement render. A few white fantail doves are in residence.

*Dating:*
This is a vernacular building of the early nineteenth century, illustrating a late stage of pigeon-keeping in which small numbers of birds were fed in the farmyard like other poultry. By that time the earlier social exclusiveness had lapsed, and there was no longer any prestige to be had simply by keeping pigeons. Larger dovecotes were still being built in extravagant architectural styles (as at Coldham Hall and Cockfield Hall), and smaller ones as ornamental garden buildings (as at Scotland Place), but here the dual-purpose building was constructed as simply as possible, re-using old materials wherever practicable. The use of the hipped roof with gablets at this late period indicates a conservative attitude on the part of the builders, who evidently decided that a design which had been used successfully for hundreds of years must be what suited the pigeons best. The large amount of re-used timber suggests that the builders dismantled a much earlier timber-framed dovecote (not necessarily at this site) and rebuilt one in similar form, but with masonry walls, using the techniques of nineteenth-century carpentry.

## HOME FARM, PETTISTREE (TM 294 549)
Home Farm is ¼-mile west of the parish church. The farmhouse has been rebuilt early in this century. The pigeon-tower is isolated in the middle of a paddock, 35 yards east-north-east of the nearest contemporary farm building (Plate 11a, opposite page 96). It comprises a pigeon-loft standing on an open lower stage, like the *colombiers sur piliers* of France, with a door facing south-south-east towards the public road, and a small aperture to the rear. The whole is 15 feet high to the eaves, with a plain-tiled pyramidal roof and a skeletal modern louver.

*The lower stage:*
The supporting stage is 9½ feet square, the fabric composed of unworked flints set in lime mortar with dressings of red brick, forming walls 13 inches thick. Each side has a round arch 6½ feet high by 4 feet wide. The bricks are 8⅞ x 4¼ x 2½ inches, of sieved and pugged clay, kiln-fired, uniform in size, laid in lime mortar.

### The pigeon-loft:

The loft is of brick, set back 3½ inches on a course of weathered stretchers. It has been rendered, but this is not an original finish, for the brickwork is of high quality, in Flemish bond, evidently intended to be seen. The doorway is 4 feet high by 2 feet wide, below a shallow segmental arch of gauged bricks. The original ledged door is present, in a renewed frame. It has external strap hinges and is fitted with a hinged bar and hasp for a padlock. The aperture to the rear is 1 foot 8 inches high by 1 foot 3 inches wide, only 1½ feet above the floor of the loft. The frame and door have been renewed.

### The nest-boxes:

The brick walls and roof are plastered inside. When examined the nest-boxes had been dismantled for repairs to storm damage above, but some of the components were still present, unfired clay tiles 9 x 9 x 1½ inches. These are not clay bats, for the fabric is of finely sieved and pugged clay, without straw or other fibre - exactly the same as floor tiles, but left unfired. Traces of mortar on the walls show how they were formerly assembled; there were 9 tiers of nest-boxes, with 34 in each complete tier. Allowing for those omitted at the front and rear apertures there were 286 nest-boxes in all. Nothing remained of the fronts of the boxes, but the Listed Building description of 1984 may suggest how the fronts were built, for it stated that the nest-boxes were of wood. This may be taken to mean that the fronts were of wood, for the other components would not have been so visible. Messent's description of 1936 adds a little to our information: 'There are nearly three hundred nesting boxes inside, these consist of horizontal timber shelves with vertical partitions of clay; each box is approximately 9 inches square'.[79] The marks on the walls confirm that the 1½-inch thick clay tiles stood on edge, supported on quite thin wooden shelves.

### The roof and louver:

When inspected these had been severely damaged in a storm, but if the repairs followed the earlier form there had been the framing of a square pipe, and an incomplete square louver. Messent's description is unhelpful: 'The usual cupola with finial crowns the roof'. Regrettably, original cupolas are not so familiar sixty years later.

### A rare feature:

Inside the doorway there is a wrought iron gate, opening inwards and secured by a hasp and staple (Figure 38). Nothing like this has been found in any other dovecote in Suffolk, nor (on present information) has anything similar been reported elsewhere. Its purpose can only be discussed speculatively. It seems to have some functional relationship with the small aperture at the rear, which is too low to be of much use as a window, but which can be reached easily by ladder (Plate 11a). One suggestion is that when removing adult pigeons two people worked together from outside, one driving them out from the front, the other catching them as they emerged at the rear. This technique could be used for culling old birds, or for taking

Plate 11a (top): The pigeon-tower at Home Farm, Pettistree, from the west.
Plate 11b (bottom): The kennels range and pigeon-loft at Claycotts, East Bergholt, from the south-west.

Plate 12a (top): The dovecote at Eriswell Hall, from the south-east.
Plate 12b (bottom): In the northern cell at Eriswell Hall, showing the roof structure, and at right the partition wall of brick-nogged timber framing, still exhibiting stains from the former nest-boxes.

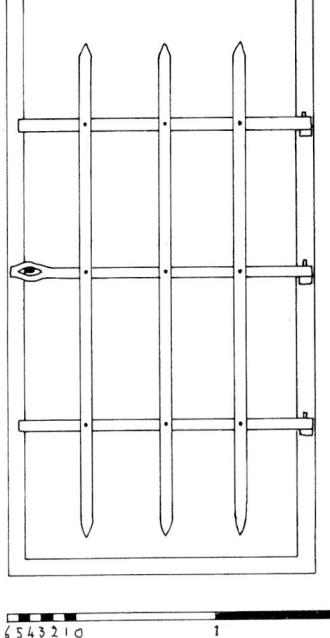

Figure 38: Measured elevation of the wrought iron gate at Home Farm, Pettistree, as seen from inside the loft. (Scale in feet)

adult pigeons for trap-shooting. Readers who have relevant experience are invited to comment on this explanation, or to offer others.

## *Dating:*

This type of pigeon-tower, supported on four masonry piers, is rare in Britain. Cooke illustrated one at Megginch Castle in Perthshire, and there is another at Chastleton in Oxfordshire which he was unable to see. In the details of design neither is similar to this one. Messent wrote that this 'is said to date back to about 1775', without stating the source of his information, and Pevsner repeated it without comment.[80] The quality of the bricks suggests that it is much later, and indeed it is Listed as nineteenth century. It is not shown on an estate map of 1800, nor on the tithe map of 1842. It is shown in the First Edition Ordnance map of 1883.[81] This confirms the impression that it is Victorian. Whether it was used to produce squabs for meat, or adult pigeons for trap-shooting, or simply for the fancy breeds, remains uncertain. As an ornamental building it seems to be in the wrong place; it would be more appropriately situated within the garden of a fashionable country house.

## *Additional information:*

Minor frost damage to the bricks accounts for the fact that it has been rendered at a later date. The softwood floor joists are original. A main joist across the middle is of 4¼ x 3 inch horizontal section. Three common joists to each side of 3 x 2⅛ inch vertical section are jointed into it. The original floor-boards are still present, with modern boards over them.

## CLAYCOTTS, FLATFORD LANE, EAST BERGHOLT (TM 077 342)

This timber-framed pigeon-loft is inserted in the series at this point because it illustrates an even later stage of pigeon-keeping. Claycotts is a mainly sixteenth-century house on the eastern outskirts of East Bergholt. Extending to the rear is a range of Victorian kennels in Gothic style; the pigeon-loft is built over them as part of an integrated design (Plate 11b, opposite page 96). It is 8 feet square, mounted 8 feet above ground, with weather-boarded walls and a four-gabled roof. The door faces west (accessible only by ladder), and there are windows to north and east.

### The frame and cladding:

The structure is simply constructed of hardwood. Some of the white-painted weather-boarding has been renewed, but where it remains undisturbed there is original brick nogging between the studs. The roof framing is of hardwood, of nailed construction, clad with hand-made plain tiles.

### The door and windows:

The door is 4 feet 10 inches high by 2 feet 2 inches wide, with an arched head of three-centred curvature which matches the three doorways below. It opens outwards, sheltered by a simple drip-ledge above. The unglazed windows in the left and rear gables are 1 foot 5 inches high by 10 inches wide, with two-centred arched heads. Originally they were protected against birds of prey by closely-spaced vertical wooden bars - most of which are now decayed. Planted outside the weather-boarding of the south gable is a false window-frame of similar shape, which may be from a real window removed when this elevation was re-boarded.

### The flight holes:

In the gable above the door are nine flight holes arranged in three rows (of one, three and five), each 6½ inches high by 4 inches wide; they have elegantly profiled heads of steep ogee curvature. Outside each row of flight holes is a shelf 4 inches wide, with a similar shelf on the inside.

### The nest-boxes:

The nest-boxes are made of ½-inch pine, and completely cover the rear and side walls. There are 6 tiers with 17 boxes in each, making 102 in all. Each box is 9 inches high by 14 inches wide by 10 inches from front to back. The planks project to form alighting ledges 4 inches wide. The entrance holes are of inverted-U shape 6 inches high by 4½ inches wide, each within half an inch of the side of the box. Vertical strips 4 inches wide have been attached between the alighting ledges, so that each pair of entrance holes is enclosed in a projecting frame. This arrangement of nest-boxes in pairs follows advice by St. John Priest in 1810 and J. C. Loudon in 1825.[82] Because the breeding cycle of pigeons is so short there are times when one pair occupies two nest-boxes, incubating eggs in one while continuing to feed squabs in the other. The interior has been treated with lime-wash many times.

The house Claycotts is Listed as Grade II\*; the pigeon loft is included in the same description. About twenty pigeons were in residence when inspected - some white fantails, some of mixed breed.

*Cassell's Household Guide* of 1870 advised its readers to keep common pigeons primarily for entertainment: 'Their variety and beauty of shape and plumage, the tenderness of their voices, their absurd antics and small vanities, their struttings and love-makings, all combine to render them an endless source of amusement and pleasure to every properly organised mind'.[83] This pigeon-loft was built about that time, when pigeons were kept mainly for delight, but also to produce delicate fare for the household.

# Chapter 8
# Three oddities

**ERISWELL HALL, ERISWELL** (TL 721 807)

This must be one of the strangest dovecotes in Britain, for it has been converted from the nave of a medieval church. It is isolated on a small eminence in otherwise flat farmland, 200 yards east-south-east of Eriswell Hall. By about 1720 the church had been abandoned, and only the unroofed nave remained standing.[84] For the conversion in 1754 the western part of the nave was demolished, new east and west walls were built of brick and timber framing, and a roof was built of characteristic dovecote type (Plate 12a, opposite page 97). The resulting building is 30 feet from north to south, 20 feet from east to west, and 17½ feet high to the eaves. The plain-tiled roof is hipped, with large gablets to north and south.

*The former church:*

The parish church of St. Peter was built of flint rubble with limestone dressings, accurately orientated east-west. What remains is the eastern third of the nave. The north side retains a tall transomed window of the fifteenth century, with a moulded four-centred arch, jambs and label of limestone, and a sill of Barnack stone (Figure 39); the rear-arch and splayed jambs are moulded too. From inside one can see the east jamb of a thirteenth-century lancet window which was blocked when this window was inserted, and the east jamb of another lancet to the west. Inside the south wall there is remaining evidence of the jambs and part of the segmental arch of a window, but the rubble fabric above and below it has been demolished. The long east side retains some original flint rubble fabric at each end, representing the reduction of the nave to meet the chancel (Plate 12a). How much else remained at the time of conversion is not known, but all unwanted masonry was demolished. In the eighteenth century there was no respect for medieval buildings.

*The conversion into a dovecote:*

The Perpendicular north window was blocked with brickwork up to the transom. Above that the tracery was cut away and a rectangular window was inserted, cutting into the arch at the top corners. The large aperture left in the south wall by the demolition of its window and the associated masonry was built up from the ground in brickwork, leaving a rectangular aperture for a high window (now boarded over, Plate 12a). In this wall is an arched stone plaque with inscribed initials which appear to be T.R., with the date 1754. A brick wall was built to close the former division between nave and chancel, with diagonal buttresses at the corners. A new timber-framed wall was built across the nave, leaving two doorways with a high window above each, and it was brick-nogged. A timber-framed partition, also nogged, was built across from east to west (up to eaves level only). This divided the space enclosed into two equal cells, each having a door and two windows.

Figure 39: The remains of a medieval church at Eriswell converted to a dovecote, from the north-west

### *The doorways and windows:*
The doorways (of which the more northerly is now blocked) face the house. They are 5 feet 7 inches high by 2 feet 6 inches wide. The sills were set at least 1 foot 9 inches above ground. The ground level has risen, so they were higher when built. The new east windows were 3 feet high by 2 feet 6 inches wide, one foot below the eaves. The window formed in the north wall has been reported already. The new window in the south elevation is 3 feet 2 inches high by 2 feet 8 inches wide. It is not in the middle of the wall, because the panel of brickwork is off-centre, but it is placed centrally in that panel.

### *The brickwork:*
The bricks are 8½ x 4¼ x 2 inches, hand-made, similar in colour to the gault bricks which were to become highly fashionable (as at Hitcham House and its dovecote), but smaller and not so finely made. They are laid in lime mortar in English garden-wall bond, comprising one course of headers to three, four or five courses of stretchers; four courses rise 10 inches. The three masonry walls have an eaves cornice consisting of one course of separate headers forming dentils, one

course of stretchers projecting one inch, and a second course of stretchers projecting 2 inches.

## *The interior:*
The partition wall is of high quality timber framing, all of hand-sawn oak, jointed and pegged, and filled with one thickness of bricks of the same quality as on the outside. There is no evidence that it extended above eaves level, so the roof was always open from end to end, as it is now (Plate 12b, opposite page 97). The date 1754 is simply inscribed 3 feet above ground in a chalk block in the east wall. This would have been covered when the nest-boxes were built, so it confirms that the conversion was done in that year.

## *The former nest-boxes:*
The nest-boxes are missing. They were mounted on a plinth 1½ feet high by 10 inches wide all round each cell, composed of stone rubble with a top layer of bricks, which remains *in situ*. Up to a height of 9 feet the walls have been lime-washed, but above that stains on the studding show that the tiers of nest-boxes were 10 inches high (Plate 12b). There is sufficient height for 19 tiers. Any calculation of the number must depend on assumptions about their width and the thickness of the vertical partitions between them. If they were 12 inches wide and made of sawn boards one inch thick there is sufficient space for 50 in each tier of each cell. Allowing for those omitted at the doors and windows there would have been up to 1,800 nest-boxes. If the vertical partitions were of a bulky material like chalk blocks there would have been substantially fewer.

## *The roof:*
The roof is built in five bays with joggled butt-purlins, tenoned and pegged. The gablets are formed by collars tenoned across the principal rafters. Joists have been set across the roof, bird-mouthed and nailed to the higher purlins of the middle bay, nailed to the common rafters in the other bays where the purlins are too low; they support a flight platform running the full length between the gablets (Plate 12b). There was formerly a pipe above each cell, but the remaining evidence is minimal - holes in the flight platform and some nails in the adjacent collars, too inaccessible to permit detailed examination or measurement. The gablets have been closed with cement render on modern lathing. The plain tiles are mostly hand-made, of various colours, with some repairs executed with machine-made tiles.

## *Later developments:*
After passing out of use as a dovecote this building was adapted as a stable or other animal housing. The more southerly of the two west doorways was enlarged, an extra window was cut in the studding above it, the more northerly door was blocked, and a new doorway was made through the partition wall, 7 feet high by 2 feet 8 inches wide. A new window was cut in the lower part of the fifteenth-century north window, which had been bricked up to the transom.

## Additional information:

No former owner with the initials T.R. has been traced, but Peter Northeast suggests that if they are read as I.R. they could represent John Robinson, who was prominent in the parish in the eighteenth century. Eriswell was enclosed in 1818.

The timber framing of the west wall and the internal partition is composed of posts 8 x 6 inches, girts tenoned and pegged to them at mid-height, and primary straight braces and studding 4 x 5 inches. The end posts of the partition wall are jowled and jointed to the tiebeam in normal assembly. A clear series of assembly marks is visible on the north side of the partition wall. Below the girt the series runs from east to west across the studs and central post. Above the girt it runs from west to east. They are chisel-cut, in Roman numerals except the 9, which is represented by an oblique Z - a representation of 9 which has been reported in sixteenth and seventeenth-century carpentry in Essex.[85] Later farm buildings have been erected to east and south of the dovecote, but it is still free-standing.

One earlier conversion of this type is on record - and has already come into this story in another connection. The dovecote at Sprowston, Norfolk, which attracted the wrath of the Kett rebels in 1549 had been converted from the chantry chapel of St. Mary Magdalene. The lawyer John Corbet bought it (jointly with Sir Robert Southwell) from the Court of Augmentations; it was a place of worship until the chantry was dissolved in 1548 (see Note 25).

## ALDER CARR FARM, CREETING ST. MARY (TM 091 553)

This dovecote is almost as strange. It began life as part of a windmill, was moved and converted into a dovecote about 1860, and was converted to a third use in 1996. It is 18 feet long, 12 feet wide, and 26 feet high to the eaves. It is timber-framed and weather-boarded, with a hipped roof with gablets, clad with clay tiles. It was built as a post-mill in 1796 on Creeting Hill, in the old parish of Creeting St. Olave's. The last miller was H. Weavers, who was still there in 1858. The lease expired in October 1860, and the mill was sold for dismantling. The buck (that is, the part of the structure which rotated on a fixed post), still in good condition but without any machinery, and without a roof or upper floors, was moved to Broughton Park Farm, Creeting St. Mary, 800 yards to the south-west. There it was mounted on brick piers and converted into a dovecote. A new roof was fitted, and the inside was furnished with nest-boxes. There it remained until December 1995, by which time it had become near-derelict (Figure 40). It was moved to another position on the same farm, and converted into a workshop for a wood turner. One tier of nest-boxes was left *in situ* to indicate what they had been like earlier, and the building was repaired and refurbished. It can be seen by anyone who patronizes the farm shop (Figure 41).

## The dovecote phase:

When in use as a dovecote the buck was situated 5 yards east of the nearest farm building (a long barn range), with its obtusely pointed front wall facing north-north-east. The base was supported 2 feet above ground on twelve brick piers. The bricks were 9 x 4½ x 2½ inches, hand-made, laid in lime mortar, each pier topped

Figure 40: The windmill buck at Alder Carr Farm, Creeting St. Mary, in a derelict condition in 1991. Drawing by Pamela McCann.

by tiles and a wooden pad 1½ inches thick, and the whole tarred. The corner posts projected below the buck and extended five courses into the piers. The original doorway to the tail of the buck continued in use, but the tail steps were removed, and the only access was by a removable ladder.

### *The nest-boxes:*

The inside of the weather-boarding was lathed and lime-plastered between the members of the timber frame to form the backs of the nest-boxes. They were built of ¾-inch softwood, nailed in place, fitted between the structural timbers. Consequently the boxes were irregular in size, but mostly about 12 inches high, varying from 12 to 18 inches in width, 10 inches from front to back, and left open at the fronts. When first examined in 1991 there were 6 tiers against the short front and back walls, each of 6 boxes; staining on the wall showed that there had been another tier below. Against each of the longer side walls there were 6 tiers each of 11 boxes, making a total of 216 nest-boxes. Traces of nesting material showed that most of them had been occupied by pigeons. The whole interior had been lime-washed.

### *Decline and decay:*

Probably the building continued in use as a hen-house even after it passed out of use for pigeons. The roof was tiled originally, but by the mid-1980s it was in bad order, and corrugated iron sheets were nailed to the top rails to preserve them.

Figure 41: The windmill buck at Alder Carr Farm, Creeting St. Mary, after conversion to a wood turner's workshop in 1996.

Some of the piers were rebuilt in modern brick. Decay and storm damage destroyed about half the roof framing. The main frame was inclining to the north-east.

### *The latest phase:*

The buck was moved on a low loader to its present position 30 yards south of the barn range, immediately inside the farm entrance. The frame was repaired where necessary. The old weather-boarding was left in place, and it was covered by new weather-boarding. The lime plaster inside was conserved and lime-washed. The roof was rebuilt and clad with new clay tiles. Floors and a stair were built, and windows were inserted. It was opened with appropriate ceremony in May 1996.

## Additional information:

I am obliged to Peter Dolman of the Suffolk Mills Group for the history of the mill. It was advertised for sale in 1796 as a 'New-erected Post-Windmill'. Another advertisement of 1801 added that it had a round-house, and that the property included a new mill house. The owner was Richard Emerson, who went bankrupt in 1803. By 1844 the miller was Robert Bixby, who died in the early 1850s. The property was put up for auction in June 1854, stating that the lease would expire in October 1860, and that unless renewed the mill could be 'taken down and removed'. It was advertised three more times, and finally was sold by auction in December 1855. The site to which the buck was moved later became Alder Carr Farm.

## PARSONAGE FARM, STONE STREET, BOXFORD (TL 964 402)

This cylindrical brick dovecote is situated 15 yards north of the house, close to the River Box (Figure 42). It is 10 feet in diameter and 12½ feet high to the eaves. It has a door to the south-west, and windows to north-west and south-east. The plain-tiled roof is conical, with a modern cupola.

It qualifies as an 'Oddity' because it is a Listed Building, officially described as eighteenth century. Most of it was built in the late 1970s. It was Listed in error by an Inspector of the Department of the Environment, who apparently believed that it was a genuine historic building. In other Districts of Suffolk there was a re-survey of Listed Buildings, beginning in 1982, to correct the many faults in the earlier schedules. Babergh District was omitted from the re-survey. Everyone who examines historic buildings in Babergh finds its List still riddled with errors and omissions, but English Heritage has dismissed representations that a re-survey is needed. In 1998 the defective List for Babergh District is still in force.

## The door and windows:

The doorway is 5 feet 7 inches by 2 feet 3 inches, with a modern softwood frame. The salvaged eighteenth-century door has two large fielded panels and ovolo mouldings, a kind of door intended only for internal use. The windows have modern sheet glass; three iron bars set in the brickwork cross inside.

## The brickwork:

The cylindrical wall is of red bricks 8½ x 4⅜ x 2⅜ inches laid in lime mortar in header bond. The plinth stands 5 courses above ground, with a plain set-back of 1½ inches. The eaves cornice comprises one course of separate headers forming dentils, and two continuous courses which overhang by 3 inches.

## The roof and cupola:

The roof structure is of machine-sawn softwood, clad with salvaged hand-made tiles (which by 1998 had developed a pleasing patina). The cupola is octagonal, made of marine plywood, with two tiers of eight flight holes of inverted-U shape. It has a conical cap of galvanized iron, with a spike finial.

Figure 42: The dovecote of Parsonage Farm, Boxford, from the south-west.

***The interior:***
Everything inside is modern. Sixteen Dragoon pigeons, an ornamental breed, were in residence when inspected.

***The explanation:***
When the present owner bought Parsonage Farm there was the ruin of a partly-demolished dovecote on site. Old photographs showed what it had looked like earlier. It had deteriorated over a long period, and a previous owner had demolished most of it in order to re-use the building materials. The architect Ray Williams, AA Dipl., RIBA, of Boxford was engaged to rebuild it with compatible materials salvaged from elsewhere, which were blended into what remained of the earlier structure. A photograph taken before the re-building began shows only an incomplete brick shell, standing 10 feet high at the north-east, but declining to ground level at the west.[86]

# Chapter 9
# Comparisons and conclusions

Suffolk has been shown to have dovecotes from all periods, and of many materials and plan types. In some cases they retain original features which have not been reported elsewhere - although it is hoped that this study will stimulate other observers to find and interpret them in other parts of Britain. This is an attempt to draw some conclusions from the survey.

## THE SITES
In 1740 an anonymous poem described the ideal site for a dovecote:
> *That Situation will your Doves delight,*
> *Where the Air's pure, and nothing stops the Sight,*
> *Then place your Cote, where from the rising Ground*
> *The Fields in Prospect open lye around.*[87]

Because pigeons are attracted to high places it might seem important to set the dovecote on the highest part of the site. Evidently this consideration was outweighed by others, for some of the dovecotes described here are downhill from the associated house. The same writer advised that the dovecote should be situated well away from woodland:
> *For oft within the Grove's deceitful Shade*
> *The Hawk his Prey is ready to invade.*

In 1698 Roger North had offered similar advice: 'Woodlands harbour haukes, the desperate enemys of these poor birds that inhabit with us . . . And the lovre should not be lower than the adjoyning buildings, and neerest trees; for the hauks will have an advantage to descend upon them, that cannot strike so well rising'.[88] None of the dovecotes described here is in the vicinity of old woodland. Where now there are tall trees nearby it can be shown by measuring the girth that they have been allowed to grow there since the dovecote passed out of use in the nineteenth century.

North wrote of pigeons: 'If they are among too much buissness and noise, they will be frighted. If retired, the hauks will be too saucy; the passing of men to and fro frights them'. That is, the dovecote should be situated far enough from other buildings to avoid disturbance, but not so far as to expose the pigeons to danger from birds of prey. The distance from the nearest contemporary building has been stated here (disregarding later buildings). With the exception of the Victorian pigeon-loft at East Bergholt, all these dovecotes were built as free-standing buildings - although the distance from other buildings varies widely.

Contemporary writers offered confusing advice about the proximity of water. Leonard Mascall wrote in 1581 that the dovecote should be 'placed very nye some water'.[89] One anonymous writer in 1735 advised that 'the pigeon-house should be placed at some small distance from water', and another in 1740 said that pigeons were disturbed by the noise of moving water. Roger North in 1698 said that 'water should be at a moderate distance, and quiet'. The apparent conflict of opinion was resolved by Daniel Girton in 1785. He agreed that pigeons needed access to water, but pointed out that they were likely to be disturbed by the noise from a mill.[90] The feral pigeons in a modern city centre tolerate a high level of ambient traffic noise, but any sudden loud noise will send them all into the air. In the quieter environment of a rural manor the sudden roar produced when a mill-sluice was opened would certainly have been sufficient to alarm the birds. When most manors had a mill it must have been difficult to find a place for the dovecote which was far enough from it to avoid disturbing the pigeons, but close enough to human activity to deter hawks. In this survey the distance from the nearest water has been stated. (At Bury St. Edmunds special circumstances applied. The Abbot's pigeon-tower was only 40 yards from the mill, but for defensive reasons both were concentrated within the perimeter wall).

Most of the post-medieval dovecotes described here are prominently displayed at the front of the house, or are visible from the approach road. The only exceptions are at Aspall, where the dovecote was raised on a mound several hundred yards away, and at Hitcham and East Bergholt, both minor dual-purpose buildings situated behind the house.

As reported in Chapter 1 in connection with Charles Waterton's new dovecote, to stock a new dovecote it had to be completed before the pigeons could be excluded from the old one. Therefore even if an ideal site were available, which met all the requirements described above, it would still have to be abandoned when the decaying or unfashionable older dovecote was replaced by a newer one.

## PLAN FORMS

There have been attempts to classify dovecotes by their plan forms, and to interpret certain shapes as characteristic of particular periods. This survey confirms that that is simplistic and unhelpful. The earliest two in the series are oblong and hexagonal (Mildenhall and Bury St. Edmunds). The round dovecotes date from the early seventeenth century and the early nineteenth century (Flixton Hall and Coldham Hall). The square buildings are distributed through four centuries. Octagonal dovecotes are sometimes assumed to be characteristic of the eighteenth century, but of those reported here Gipping was built in the seventeenth century, and Great Glemham and Letheringham were built in the early nineteenth century.

## MATERIALS

Similarly, it has been suggested that the materials used determined the plan shapes. Timber framing may lend itself conveniently to forming buildings of rectangular shape, but the skilled carpenters of Suffolk were capable of building to much more complex plans - as may be seen on the irregular sites in the towns.[91] At Gipping they built an octagonal dovecote, and at Aspall an hexagonal one. Of the fourteen surviving dovecotes mainly constructed of brick, three are round, four are octagonal, five are square and two are oblong. This demonstrates that the general appearance was determined by the client; the master builder then applied his art to construct what his client wanted. From the eighteenth century dovecotes for important clients were designed by architects. Typical examples are 'South Suffolk B', Kentwell Hall, Great Glemham, Letheringham, Hitcham House, Coldham Hall, Cockfield Hall and Pettistree.

## DOORWAYS

An eighteenth-century writer said that the door should be visible from the house 'because the master of the family may keep in awe those that go in and come out of the pigeon house'.[92] The reports state whether the door was visible from the house. Most of them were, but three were not (Flixton Hall, Kentwell Hall, and Polstead). In size the doorways fall into two groups: many of the early ones were only 4 to 4½ feet high (although some were enlarged later); from the eighteenth century most are of normal domestic size. Two reasons have been given for keeping the doorway small: (1) the pigeon-keeper could block the aperture with his stooping body to prevent birds escaping as he entered, (2) the door occupied wall space which otherwise could be devoted to nest-boxes. Taller doors were adopted in the eighteenth century because their proportions suited the architectural requirements of the time.

Some of the original doorways are surprisingly wide - 3 feet or more at Gipping, Exning, Ousden and Kentwell. This may suggest that the builders were allowing enough width to use a barrow or handcart when removing the valuable manure.

## WINDOWS

Contemporary writers confidently offered advice about the windows of dovecotes, but it differed considerably. For instance, in 1581 Leonard Mascall said that the single window should face south for warmth. The anonymous author of 'The Sportsman's Dictionary' of 1735 agreed, and stressed that the window should not face east.[93] In 1746 a correspondent of *The Gentleman's Magazine* described a round dovecote without windows where the pigeons had never prospered, and said that he had improved it by inserting windows: 'Four oval holes being made at equal distances from each other, about two foot high, and one foot wide, and about eight feet from the ground, the pigeons immediately took more kindly to the house, and have thrived and increased since that time'.[94]

This survey has found great diversity both in the number of windows provided, and in the directions they faced. Nine of these buildings had no windows originally, relying on the illumination which filtered through the pigeon entrance in the roof. Five had one window, nine had two windows, seven had more. Where there was only one window, in three cases it faced south (as advised by the writers quoted above), but one faced west, and (contrary to that advice) one faced east. Where two windows were provided, in most cases they were opposite each other; but at 'South Suffolk A' the windows faced south and east, and at Goswold Hall they faced north and east. Evidently the decisions on how many windows were required, which way they should face, or whether they were necessary at all, were very much a matter of personal opinion.

What was never in doubt was that windows had to be protected against birds of prey. The correspondent of *The Gentleman's Magazine* quoted above continued: 'Wire-lattice was nail'd before all these holes, in order to keep vermin out' (in his time birds of prey were simply regarded as vermin, though we value them today). The remains of rusted wire grills have been reported at Great Glemham and Letheringham, and iron bars remain *in situ* at Flixton Hall, Great Thurlow and Coldham Hall. Holes where they were formerly fitted remain elsewhere. The windows of the pigeon-tower at Hitcham House were protected by cast iron grills until the recent restoration. Closely-spaced wooden slats were common elsewhere; they survive in good order at Flixton Hall, Gipping and Cockfield Hall. Flixton dovecote still retains one glazed window typical of the seventeenth century, and another made to a similar design in the eighteenth century, both complete with leaded glazing.

There is documentary evidence of glazed windows in dovecotes of high status at a period when most farmhouses still had only unglazed windows. In 1441 Peterhouse College, Cambridge, spent 22d. on glazing one window of its great dovecote at Thriplow. The dovecote of Queen's College had at least 13 square feet of glass in 1538; and that of Jesus College had at least 44 square feet in 1576.[95] By the last quarter of the sixteenth century the price of window glass was falling steeply, and it was being widely adopted in farmhouses, so we need not dismiss the physical evidence of small glazed windows in the relatively minor dovecotes at Church Farm, Clare, and 'South Suffolk A'.[96]

Six of the dovecotes recorded originally had a window immediately above the door. This design simplified the masons' work, for one aperture in the masonry could serve both purposes, but they are found in timber-framed buildings too. Perhaps there was a more functional reason: the pigeon-keeper would want to glance inside to see if the pigeons were settled, or if a bird of prey was trapped inside, before he opened the door.

## ROOFS

One clear conclusion is that Suffolk has a strong regional tradition in the roofing of dovecotes. Eight examples have been found of the hipped roof with open gablets, which is uncommon beyond East Anglia and Essex. There is no consistency in how the gablets were orientated. Dovecotes of more architectural designs in the eighteenth and nineteenth centuries departed from this tradition and adopted pyramidal or conical roofs, but a hipped roof with gablets was built in 1754 at Eriswell, and another later still at Drinkstone. Elsewhere in England and Wales four-gabled roofs are common on square dovecotes - an ingenious design which provides shelter from wind, and inclined surfaces facing the sun in all conditions.[97] Only one example has been found in Suffolk, on the Victorian pigeon-loft at East Bergholt. The only dovecote which had a simple ridged roof is that at Goswold

Hall, where it was determined by the decision to embellish it with curvilinear gables (Plate 3b, opposite page 58). Even there, the high gables provided enough shelter from wind to form a warm micro-climate on the roof. Evidently there was no disagreement about pigeons' need for sheltered slopes on which to perch, but various owners provided it in different ways.

## PERCHING LEDGES

There is little or no evidence of external perching ledges in Suffolk. Two dovecotes have projecting brick courses at mid-height (Kentwell Hall and Coldham Hall, Figure 24 and Plate 9a, opposite page 88), but many domestic and public buildings have similar string courses provided for purely visual reasons. They are as wide as some internal alighting ledges, so the pigeons would have been able to perch on them, but they may not have been intended for that use.

On timber-framed dovecotes perching ledges were simply planks supported on nailed brackets; the zoologist J. Whitaker aptly described them as *sunning boards*.[98] When the nails rusted away the planks would fall. They would not have been renewed when the buildings ceased to be used as working dovecotes, so it is possible that some Suffolk dovecotes had wooden ledges earlier. However, in England perching ledges are most often found at exposed sites near the coast; they are very common in the more turbulent climate of Scotland. Apparently they were not really necessary in the mild climate of Suffolk.[99]

One should not forget that other buildings in the complex provided additional perching space. Roofs of barns, or the house itself, would be in sunlight at different times of day, and sheltered against wind from various directions. Simple observation shows that pigeons do not always choose to perch on the buildings provided for them; they can often be seen on the roofs of neighbouring buildings (as shown at the left edge of Plate 11b, opposite page 96).

## LOUVERS, CUPOLAS AND LANTERNS

The main purpose of the louver - whatever its form - was to keep out birds of prey, while allowing the pigeons to enter and leave freely. Another function - very necessary when there were many dovecotes - was to identify the building clearly to the pigeons, and to guide them home in poor visibility. Original louvers or cupolas are rare in Suffolk, as they are elsewhere. Although the square turret at Ousden Hall has been boarded over it may have remained unaltered in other respects (Figure 37). The only others which are worth considering as historical evidence of working practice are the late examples at Hitcham House and Cockfield Hall (Figures 28 and 31), and the disintegrating nineteenth-century lantern at Flixton Hall (Plate 3a, opposite page 58). Without scaffolding it is impossible to examine them closely, to determine how much of the structure is original.

When dovecotes passed out of economic use they were neglected until this century. Most of the cupolas and lanterns which one sees today are modern reconstructions, in some cases based more on imagination than on reliable evidence of what was present originally. Occasionally old photographs or drawings survive in public archives, or in private hands, recording what was there earlier. Close examination of the nails used may determine whether a repaired or reconstructed louver is authentic, for hand-made nails were being superseded by machine-made nails in the late nineteenth century.[100] The small flight holes at Polstead, Drinkstone and East Bergholt are characteristic of very late, or modern pigeon-keeping. They would be effective in keeping out birds of prey, but they would not allow large numbers of pigeons to take refuge simultaneously when danger threatened.

## PROTECTIVE DEVICES AGAINST BIRDS OF PREY INSIDE THE BUILDING

Further protection, particularly against sparrowhawks, was achieved by the use of pipes descending into the interior. It is pleasing to find that four pipes have survived in good order (at Great Thurlow, 'South Suffolk A', Polstead and Coldham Hall), and that there is fragmentary evidence of others. The only other regional study which has taken notice of them is Beth Davis's survey of South Cambridgeshire. She described and illustrated two (and residual evidence of others), which she called 'chutes'; their protective purpose was not discussed.[101] They certainly exist elsewhere, but they have not attracted comment in the literature.[102] One suspects that in some cases the remaining evidence of a former pipe has been destroyed by well-meaning restorers who failed to recognize its significance.

The inserted entrance loft and pipe at Coldham Hall is a rare or possibly unique construction.

At Clare and Great Glemham special care has been taken to deny any refuge to a bird of prey which managed to penetrate to the interior. In the former the upper surface of a free-standing tie-beam has been blocked by studding and wattle and daub, and the arrises have been deeply chamfered to avoid leaving a horizontal surface (Figure 11). At Great Glemham the tops of the brick nest-boxes have been finished to a steeply inclined surface, for the same reason (Figure 26). Roger North was aware of the problem, for in 1698 he wrote: 'that winged vermin might have no means to fly out, if once ventured in, we could not have any cross-girders'.[103] In 1810 the Reverend St. John Priest criticised a dovecote at Shardlow, Buckinghamshire: 'The pillar in the middle of the house should not reach too near the cupola, lest it should serve as a resting place for hawks, owls, and other enemies of pigeons'.[104]

One wonders what measures were taken elsewhere. Probably clay daub was used in a similar way to block the space between the highest tier of nest-boxes and the roof. This was common practice in other buildings; in contemporary accounts it was called 'beam-filling'. It would tend to be lost during roof repairs. But what - if anything - was done to prevent birds of prey from perching on the horizontal beams which supported the upper bearing of the potence? Contemporary writers did not discuss the problem, and no satisfactory physical evidence has been found. Was clay daub used there too?

## THE PROVISION OF NESTING PLACES

Nest-holes incorporated in the structural fabric have been reported at Mildenhall, Bury St. Edmunds, Badley Hall, Stoke-by-Clare, Goswold Hall, Kentwell Hall and Cockfield Hall. Elsewhere nest-boxes have been found made of oak, pine, wattle and daub, bricks, clay bats, and other forms of unfired clay. In some former dovecotes the nest-boxes have been removed, but sufficient evidence remains to allow deductions to be made about their size and number. The secure totals range from 1,178 nest-boxes at Flixton Hall to 102 at East Bergholt. More speculative calculations suggest that there may have been 1,100 - 1,300 nest-holes at Mildenhall, and up to 1,800 nest-boxes at Eriswell. This is within the national range. The largest dovecotes in Britain have about 2,000 nesting places, but the great majority of traditional dovecotes have between 1,000 and 300. Anything less must indicate a very small estate, or the late stage of pigeon-keeping in which the birds were fed in the yard like other domestic poultry.

One cannot estimate the number of pigeons in a flock simply by multiplying the number of nesting places by two. As noted at East Bergholt, one breeding pair needed to occupy two nest-boxes when the female produced a new clutch of eggs before the previous squabs had reached maturity. Probably some were always left vacant. The number of male birds did not always match the number of females, although the pigeon-keeper tried to achieve that balance by culling surplus males.

It is clear that there was no consensus among pigeon-keepers about what was the best size or shape of nesting place, for they vary greatly in every dimension. Elsewhere in the world there is more uniformity. For instance, in the highly diverse pigeon-towers of Iran it is reported that the nest-holes are always 8 x 8 x 10½ inches, perhaps determined by the mud bricks used in their construction.[105] No satisfactory explanation has been found for the diversity of shapes and sizes in Suffolk, except that different owners and pigeon-keepers had different opinions on what was best for their birds.

At Kentwell Hall the builder provided a raised course of bricks across the fronts of the nest-holes, presumably to prevent the eggs or squabs from falling out (Plate 7). At Badley Hall the entrance-holes were cut slightly above the base-boards to provide similar raised lips (Figure 5). At Clare the clay bats in the nest-boxes were deeply dished, probably for the same reason. Other pigeon-keepers saw no necessity for this extra protection. The activity of pigeon-keeping seems to have offered unlimited scope for diversity of opinion.

(In addition to the buildings described here there are surviving nest-boxes in pigeon-lofts which have been introduced into the roofs of older buildings. Edward Martin has reported one example at Park Farm, Lavenham. They represent a late and small-scale phase of pigeon-keeping. It is not practicable to collect descriptions here).

## THE USE OF UNFIRED CLAY

In Suffolk, more than elsewhere, there was a vigorous process of experimentation in the use of unfired clay for nest-boxes. One influence was the search for substitute materials following the introduction of the Brick Taxes in 1784 (tiles were exempted in 1833, and the tax was abolished in 1850). Unusual products made of unfired clay have been reported at Gipping (Plate 1b), Aspall Hall and Pettistree. Clay bats, which are familiar in dovecotes in Cambridgeshire and north Essex, have been reported at Clare, Scotland Place (Figure 25) and Coldham Hall (Figure 30).

Another influence was the belief that pigeons preferred the softer, 'warmer' materials. This view was expressed by several writers from the sixteenth century onwards.[106] In the eighteenth-century brick dovecote at 'South Suffolk B', where the external design was determined by the fashionable style of the day, the nest-boxes are made of clay daub moulded on coppice sticks. The high quality of the brickwork suggests that no expense was spared, so it seems likely that they express the owner's belief that pigeons preferred this material. The same belief was clearly expressed at Goswold Hall and Kentwell Hall, where brick nest-boxes have been carefully lined with clay daub. It is equally significant that at other dovecotes the brick nest-boxes are not lined with clay (at Stoke-by-Clare, Flixton Hall, Great Glemham, Letheringham and Cockfield Hall). Evidently the belief that pigeons preferred nest-boxes of soft materials was not universally shared.

It is worth noting that true clay lump has not been found at all. In Norfolk clay lump on a brick base was considered strong enough for building windmills, and at least one dovecote built of clay lump has been found there.[107] The common belief that clay lump is an ancient material in East Anglia is a fallacy. Few standing buildings of clay lump are earlier than the 1830s; contemporary writers described it as a new technology until the 1850s.[108] By the time construction in clay lump had become common in Suffolk, few dovecotes were being built.

## INTERNAL LEDGES

Alighting ledges or steps were not essential, for some medieval dovecotes do not have them at all. Elsewhere in the country it is quite common to find an alighting ledge provided to every second or third tier of nesting places. Quite why they were so generally adopted in post-medieval dovecotes has not been satisfactorily explained. All the dovecotes described here had an alighting ledge or step to each nesting place. Most of them are between 1½ and 3 inches wide, which was sufficient for pigeons but deliberately difficult for birds of prey. The wooden ledges at East Bergholt are 4 inches wide, and the ledges of wattle and daub at 'South Suffolk A' are 5 inches wide. At Flixton Hall the builder changed the design after completing the first three tiers, for up to that level he provided a continuous ledge to each tier, but above that he provided a separate step to each nest-box (Figure 21).

## POTENCES OR REVOLVING LADDERS

Suffolk is fortunate in having three dovecotes with potences in good order (Flixton Hall, Kentwell Hall and Cockfield Hall). In addition there are substantial remains of potences at 'South Suffolk B' and Coldham Hall, and some residual evidence at Gipping and Ousden. They occur in round, octagonal and square dovecotes. Cooke suggested that a potence would not have been of much use in a square dovecote, but the evidence contradicts him.[109] A practical demonstration at the square brick dovecote at Kentwell Hall established that there was no difficulty in reaching into the corners from the ladder. Potences are common in square dovecotes in Scotland.[110]

At Great Glemham and Letheringham dovecotes which are octagonal outside have nest-boxes arranged round a cylindrical curve, as if to facilitate the use of a potence. If potences were formerly present there the evidence has been destroyed by later uses. The same shape occurs at Crowfield Hall; both there and at Aspall Hall we have oral evidence that potences were formerly present. Therefore it seems likely that formerly they were common - even where no trace survives now.

So far no reliable evidence of a potence has been produced earlier than the one Roger North designed at Rougham, Norfolk, in 1698 (which he called a 'revolving ladder').[111] Where they are present in much earlier buildings it is sometimes apparent that they have been installed after the original construction. The clue to look for is whether the beam or beams supporting the upper bearing show signs of having been inserted since construction; often they blocked existing nest-boxes.

## OTHER INTERNAL FEATURES

At Cockfield Hall there is a central brick structure which provides an additional 48 nest-holes. Other examples are known; this feature may have been common at one time, but in most cases it has been removed to clear the floor for a secondary use. At Flixton Hall a central plinth of similar size was built purely to support the potence, and does not provide extra nest-holes.

Also at Flixton Hall an exceptionally large feeding platform is mounted on the axis of the potence (Plate 4, opposite page 59); and there is residual evidence of a smaller one at 'South Suffolk B'.

## PROTECTION FROM BROWN RATS

Dovecotes which were built before the middle of the eighteenth century of hard materials, on deep foundations, required little alteration when brown rats were introduced. At Flixton Hall the only change required was to rebuild the doorway at a higher level, above the reach of rats. Other doors which are now missing may have been protected against gnawing by iron plates - or by the equally effective practice of kennelling a dog nearby.

There were greater inherent difficulties with the timber-framed dovecotes built earlier; they were easily penetrated by brown rats, and their foundations were often so shallow as to be easily undermined. Common responses to the new hazard were to fill the frame with brick nogging, to remove the lower tiers of nest-boxes, or to rebuild them at a higher level. At Great Thurlow, Pakenham, and Clare the nest-boxes are missing, but at 'South Suffolk A' the frame was brick-nogged to a height of four feet, and the nest-boxes were rebuilt above that level (Figure 7). At Badley Hall an exceptional solution was adopted, lining the nest-boxes with paving bricks, but this was possible only because they were unusually capacious. The most thorough adaptation to defeat the new predators was at Gipping, where a complete inner structure of brick nest-boxes was built inside the indefensible timber structure, with a raised doorway (Plate 1b). Probably many other timber-framed dovecotes became useless when brown rats reached the area, and were abandoned or demolished. Elsewhere in the country there is evidence that some early timber-framed dovecotes were sheathed with brickwork in the eighteenth century, either wholly, or to a height of four to six feet.[112]

All dovecotes built after the middle of the eighteenth century were designed to be resistant to brown rats - either by constructing them of hard materials on deep foundations (as at Kentwell Hall, Great Glemham, Letheringham, Coldham Hall, and Cockfield Hall), or by forming the accommodation for pigeons in an upper storey, supported on unclimbable walls (as at Scotland Place, Hitcham, Drinkstone and Pettistree). At Eriswell in 1754 the timber-framed walls were brick-nogged, and the doorways were set high to raise them above the reach of rats. At East Bergholt the pigeon-loft was built over kennels; the proximity of dogs would have deterred the rats, but in addition the timber frame was infilled with brick. In Cambridgeshire, timber-framed dovecotes built after the middle of the eighteenth century were raised four feet above ground on brick piers or base walls.[113]

## CONVERSION TO SECONDARY USES

Where a later use can be identified with certainty the commonest use was as a stable. Crowfield is reported to have been used for cattle. Clare and Gipping have been used for poultry. Letheringham became a gig-house, and later a garage; and the dovecote of similar design at Great Glemham became a game larder, and later a pump-house. Aspall became an apple store. The medieval pigeon-tower at Bury St. Edmunds had become a gazebo by the eighteenth century. The dovecotes at Flixton Hall, 'South Suffolk A', 'South Suffolk B', Coldham Hall and Cockfield Hall have never been converted to secondary uses, and that may be true of the derelict dovecote at Goswold Hall too. The upper storeys of pigeon-towers have remained unaltered - as at Stoke-by-Clare, Scotland Place, Drinkstone, Pettistree and East Bergholt.

Elsewhere in the country one of the commonest ways of adapting a dovecote to changed conditions in the nineteenth century was to insert a floor at mid-height, retaining the upper part as a pigeon-loft while putting the lower storey to another use. Only Clare, Letheringham and Polstead still have inserted floors. The last is an untypical conversion because the upper storey is accessible from a short ramp, and seems to have become a granary.

## DISTRIBUTION

Comparison with Arthur Young's map of the main agricultural regions of Suffolk shows that dovecotes survive in all of them.[114] There is evidence from documents, pictures, maps and field names of former dovecotes in many places where they no longer exist.[115] Those which survive represent a tiny proportion of the many which existed earlier, so it would be unjustifiable to draw any conclusion from the present distribution.

## THE DATING OF SUFFOLK DOVECOTES

The earliest examples (at Mildenhall and Bury St. Edmunds) can only be dated archaeologically. Others described here are dated in exactly the same way as other historic buildings - by close examination of their materials, structures, and styles, supplemented where possible from inscriptions, heraldry or maps.

One surprise which has emerged from this survey is how late dovecotes continued to be built in Suffolk. The county reports to the Board of Agriculture show that elsewhere in the country the keeping of pigeons on a large scale was in decline from 1794.[116] By the first decade of the nineteenth century it was being severely discouraged. For example, in 1807 Thomas Rudge wrote of pigeons in Gloucestershire: 'It is probable that the mischief they do in the fields would not be compensated by the profit, if the loss occasioned by their maintenance was to fall on the owner alone'. In 1810 in Buckinghamshire the Reverend St. John Priest wrote: 'Their depredations are sometimes so great, that no one who wishes to be upon good terms with his neighbours ought to keep them, unless they are so situated upon a farm, that they can feed only upon the lands belonging to that farm, or upon those over which the proprietors of the pigeons can claim a right'.[117] In Suffolk the evidence reported here indicates that five new dovecotes were built between 1800 and 1840 - at Hitcham, Great Glemham, Letheringham, Coldham Hall and Cockfield Hall. Only the first of these can be described as minor; the last two are major buildings each with over 600 nesting places. The small pigeon-lofts at Pettistree and East Bergholt were built even later.

It is difficult to account for this phenomenon with any certainty, but the state of agriculture in Suffolk seems the most relevant factor. Agricultural 'improvement' came late to the Strong Loam region which makes up the largest part of the county. Traditionally this region was devoted to mixed farming, more for meat and the dairy than corn. This condition persisted well into the nineteenth century. The conversion of this heavy land to arable farming was well recorded by Arthur Young, William and Hugh Raynbird and others. Much of it had been regarded as unsuitable for cereal cultivation because it was cold, wet, and difficult to work. It could be converted to arable only after deep draining, which was an expensive operation. Draining had been practised earlier on a minor scale, but the Raynbirds and their local correspondents showed that vast acreages were fully drained for the first time between 1804 and 1846. At that period very little of it was hollow-tile drainage; the typical Suffolk drain was a V-shaped trench 2½ feet deep, the bottom packed with furze and hawthorn, and covered by straw. By 1854 High Suffolk had become mainly a cereal and root farming region, but twenty or thirty years earlier much of it was still permanent pasture.[118] Pigeons could do no harm on the pasture, meadow and heathland which preceded major drainage. Dovecotes continued to be built on this 'unimproved' land until it was converted to arable, although within a generation they were to become redundant.

# Chapter 10
# What happened to Suffolk's dovecotes?

There is plenty of evidence of Suffolk dovecotes which have disappeared, but it rarely provides much detail. A very large one at Little Haugh Hall, Norton (TL 953 667), is shown prominently in a painting by Peter Tillemans of about 1720, now at the Castle Museum, Norwich. Close examination shows that it was rectangular, apparently of brick, with a fully hipped roof and a louver on the ridge; pigeons are shown perching on the roof and flying overhead. It was 120 yards north-east of the house, beyond a group of farm buildings. An undated account of about the same time records a payment of ten shillings for 'gilding the Ball of ye Dovehouse'.[119] Two dovecotes at Nettlestead were photographed in 1912 by the Reverend Edmund Farrer, at Nettlestead Chace and at Water Run Farm; the former is also shown in a painting by F. G. Cotman dated 1912, now at Christchurch Mansion Museum, Ipswich. Both had hipped gablet roofs.[120] Pevsner mentioned the one at Nettlestead Chace in 1961. He also reported (1) a square brick dovecote erected in 1600 at Stoke Ash Rectory, and (2) 'the largest dovecote in Suffolk, brick, octagonal, with a high-pitched roof and a glazed lantern' at Hurts Hall, Saxmundham. These reports were repeated in the second edition of 1974, but this cannot be taken as firm evidence that the buildings were still extant then.[121] None of them survived to be Listed in the major re-survey which began in 1982; it is unlikely that they would have been overlooked.

The dovecote at Henham Hall is shown in Figure 43. Up to parapet level it was similar to the one at Cockfield Hall, Yoxford (Figure 31). Another photograph in the same archive shows the interior and roof structure, with crossed beams supporting the upper bearing of an incomplete potence. The associated house was designed by James Wyatt in the 1790s, but it is not known whether he designed the dovecote, for the style is very different. Both have been demolished.[122] At Benacre Park, Benacre, the schedule of Listed Buildings describes a summer-house as having been 'converted in the late nineteenth century from another building, possibly a dovecote'. Only the lower part of the original structure remains.[123]

At Crows Hall, Debenham, a square brick dovecote which had already been used as a stable has been converted to a cottage (TM 192 629). Some ornamental diapering remains in the brickwork, similar to that on the main house and its associated buildings, so apparently it too dates from the sixteenth century. It was a much taller building originally; the upper part has been taken down, and windows and a floor have been inserted. Nothing survives inside of its first use.[124] At Letheringham Old Hall there are archaeological remains to a little above ground level of what was probably an early dovecote (TM 280 580). In both cases the site is remarkably similar to that of the surviving dovecote at Kentwell Hall - immediately outside the moat, and near one corner of it. At Letheringham, significantly, it is as far as possible from the surviving mill - a source of noise which would have disturbed the pigeons. At Park Farm, Thorington, the roofless ruin of a cylindrical dovecote was still visible in 1983 (TM 422 729); it has been absorbed into a mineral extraction site.[125]

Most dovecotes disappeared without record. If they could not be adapted easily to another use they were left vacant and unrepaired. In some cases the roof tiles were stripped off for re-use elsewhere, or they were damaged in storms. Without a watertight roof the building would deteriorate rapidly; eventually the remains were cleared away. By chance we have an account of the last days of a dovecote at Gazeley in 1840. In the archives of the Diocese of Ely there is an application: 'There is in the Garden part of the Glebe belonging to the said Vicarage on the North side of the said Vicarage and about 86 feet therefrom a certain ancient Dovehouse built of Studwork and Clay and having the roof thereof thatched with Straw - that such Dovehouse is in so ruinous a state as to be in danger of falling and incapable of repair and no Pigeons resorting thereto the same has become an unsightly Incumbrance'. The vicar, George Howes, was granted a faculty to demolish it, and to use the materials to repair other buildings belonging to the vicarage.[126]

Figure 43: The former dovecote of Henham Hall, Henham. RCHME, © Crown Copyright (by courtesy of the National Monuments Record Centre, Swindon).

# Chapter 11
# The future

This survey may throw some light on the problems of conservation. Of the more ancient dovecotes described here, the two which have survived in least altered condition are the late fifteenth-century pigeon-tower at Stoke-by-Clare, and the early seventeenth-century dovecote at Flixton Hall. They have not been converted to secondary purposes, and they have been repaired just enough to keep them in weatherproof condition. As William Morris observed 120 years ago, historic buildings thrive best on minimal repairs and enlightened neglect.

Deliberate restoration, with the aim of returning the building to an earlier condition, is a hazardous enterprise. Extensive works may be put in hand, sometimes at massive cost, only to find later that they are based on a misunderstanding of the historical evidence. Often the money runs out, or the programme is cut short by misfortune. Enlightened planning authorities are justifiably cautious about such proposals, and tend to use their influence (and financial aid) in favour of modest repairs, intended to keep out the weather and to prevent structural collapse.

Conversion to other uses tends to be destructive. Where little remains except an outer shell a proposal for 'change of use' may suggest the possibility of retaining that remnant. More often it proves to be simply a means to establish a new development in a place where it would not be permitted otherwise. Where the building remains substantially in good order, and particularly where internal features survive, conversion to other uses is always destructive. Residential conversion is inevitably the most destructive. The insertion of floors and stairs, of windows and dormers, of thermal insulation and modern services, often leaves the original building altered beyond recognition. The least harmful use of a surviving dovecote - if it cannot continue to house pigeons - is as general storage.

For those who need to put repairs to a dovecote in hand, some advice has been published in the *Journal of Architectural Conservation*, intended primarily for architects and others professionally concerned.[127] In Suffolk the survival of clay nest-boxes presents a special problem. They are fragile compared with other materials, and are particularly vulnerable to damage by builders. It is common to find that the highest tiers have been damaged by ladders carelessly leant against them. Those who commission repair work on dovecotes with clay nest-boxes should take special care to specify internal scaffolding or other measures to protect these vulnerable features. In recent years buildings of unfired clay, and modern ways of repairing them, have been explored by the East Anglian Earth Buildings Group, which can provide informed advice and experienced practitioners.[128]

Suffolk has fewer dovecotes than many counties. The loss of any of them, whether by neglect, destructive conversion, or unsuitable repairs, is that much more serious to those who value the physical survivals of the past.

# Notes

1. Donald Smith, *Pigeon Cotes and Dove Houses of Essex*, London, 1931.
   Peter Jeevar, *Dovecots of Cambridgeshire*, Cambridge, 1977.
   E. M. Davis, 'Dovecotes of South Cambridgeshire', *Proc. Cambridge Antiquarian Society* 75, 1986, 67-89.
   C. J. W. Messent, 'The Old Cottages and Farm-houses of Suffolk', *Proc. Suffolk Institute of Archaeology* 21, 1936, 256-60.
2. Arthur O. Cooke, *A Book of Dovecotes*, London, Edinburgh and Boston, 1920.
3. R. S. Ferguson, 'Culverhouses', *Archaeological Journal* 44, 1887, 105-16.
   R. S. Ferguson, 'Pigeon Houses in Cumberland', *Trans. Cumberland and Westmorland Antiquarian and Archaeological Society* 9, 1888, 412-34.
   Alfred Watkins, 'Herefordshire Pigeon Houses', *Trans. Woolhope Naturalists' Field Club* 1890, 9-23.
   Mildred Berkeley, 'The Dovecotes of Worcestershire', *Reports and Papers of the Associated Architectural Societies* 28, part 1, 1905-6, 332-49.
4. For instance, Cooke described a dovecote at Freswick as on the island of Stroma (as Note 2, 275). It is on the mainland.
5. Propositions put into circulation by Cooke are critically examined in J. McCann, 'An Historical Enquiry into the Design and Use of Dovecotes', *Trans. Ancient Monuments Society* 35, 1991, 89-160, and 'Dovecotes: an Addendum', *Trans. Ancient Monuments Society* 36, 1992, 137-8.
6. J. C. Pridham *et al*, *Dovecotes and Pigeon Cotes in Worcestershire*, Hereford and Worcester County Council, 1974.
   Philip Ariss, 'The Dovecotes of Gloucestershire', *Journal of the Historic Farm Buildings Group* 6, 1992, 3-34.
   I. R. Stainburn, *Dovecotes and Pigeon Lofts in Herefordshire*, Hereford and Worcester County Council, 1979.
   John Severn, *Dovecotes of Nottinghamshire*, Newark, 1986.
7. Smith, 1931 (as Note 1).
   Donald Smith, 'More Essex Dovecotes', *Essex Review* 1932, 178-80; 1933, 132-5; 1934, 97-100; 1935, 229-32.
   J. McCann and K. Robins, 'The Dovecote at Berechurch Hall, Colchester', *Essex Archaeology and History* 25 (third series), 1994, 285-8.
8. C. A. Hewett, 'Timber Building in Essex: some evidence of the possible origin of the lap-dovetail', *Transactions of the Ancient Monuments Society* 9, 1961, 33-56.
9. J. M. Ridgard (ed.), *The Household Book of Dame Alice de Bryene*, Suffolk Institute of Archaeology, 1984.
10. L. M. Munby (ed.), *Early Stuart Household Accounts*, Hertfordshire Record Society, 1986, 5-62 (accounts of Hatfield House, Hertfordshire).
11. McCann, 1991 (as Note 5), 92-6.
12. R. A. Irwin (ed.), *Letters of Charles Waterton*, London, 1955, 59.
13. Leonard Mascall, *The Husbandly Ordring and Governmente of Poultrie*, London, 1581, unpaginated.
14. Edmund Pognon (ed.), *Les Très Riches Heures de Duc de Berry*, Liber, Paris, undated, 'February', 19.
15. J. C. Loudon, *An Encyclopaedia of Agriculture*, London, 1825, 1049.
16. T. Rudge, *General View of the Agriculture of Gloucestershire*, London, 1807, 326.
17. Mascall, 1581 (as Note 13).
18. John Moore, *Columbarium: or, the Pigeon-House*, London, 1735, 14-5.
19. Thomas Tusser, *Five Hundred Pointes of Good Husbandrie*, English Dialect Society edition, London, 1878, 72.
20. M. S. Giuseppi, *Victoria County History of Surrey* 2, 1920, 306-9.
    Moore, 1735 (as Note 18), 24.
21. As translated by Charles Waterton, *Essays on Natural History*, London, 1839, 244.
22. Columella, *On Agriculture* (translated by E. S. Forster and E. H. Heffner), 363. (Book VIII, viii, 3-6).
23. St. J. Priest, *General View of the Agriculture of Buckinghamshire*, London, 1810, 39.
24. Varro: *On Farming, Rerum Rusticarum*, translated by L. Storr-Best (London, 1912), 282.
    Columella (as Note 22).
25. F. W. Russell, *Kett's Rebellion in Norfolk*, London, 1859, 31, 50.
    A. Fletcher, *Tudor Rebellions*, London, 1973, 72, 75, 142-3.

26  McCann, 1991 (as Note 5), 99-100.
27  C. Northcote Parkinson, *Parkinson's Law, or the Pursuit of Progress*, London, 1958, 89-100.
28  This matter is fully discussed in a forthcoming paper, J. McCann, 'Dovecotes and Pigeons in English Law'.
29  *The English Reports*, London, 1908, 81, King's Bench Division VIII, 620-1, 638-40.
30  Gilbert White, *The Natural History of Selborne*, London, 1788, letter 44 of 30 November 1780.
31  D. Goodwin, 'Behaviour', in Michael Abs (ed.), *Physiology and Behaviour of the Pigeon*, London, 1983, 293.
32  H. Gill and E. L. Guilford (eds.), *The Rector's Book, Clayworth, Notts,* Nottingham, 1910, 37 (the passage is quoted in full in McCann, 1991, 138).
33  H. Colvin and J. Newman (eds.), *Of Building: Roger North's Writings on Architecture*, Oxford, 1981, 100-3.
34  *Coriolanus*, 1607, V, vi, 115.
35  McCann, 1991 (as Note 5), 128-33.
36  Varro, 1912 (as Note 24), 282.
    Elizabeth Beazley, 'The Pigeon Towers of Isfahan', *Iran, Journal of the British Institute of Persian Studies* 4, 1966, 107.
37  P.R.O. sc6/869/8. For this information I am grateful to Dr. Mark Bailey.
38  McCann, 1991 (as Note 5), 101-17.
39  J. McCann, 'The Influence of Rodents on the Design and Construction of Farm Buildings in Britain, to the Mid-Nineteenth Century', *Journal of the Historic Farm Buildings Group* 10, 1996, 1-10.
40  McCann, 1996 (as Note 39), 10-11.
41  Anon., *The Sportsman's Dictionary*, London, 1735, unpaginated, under alphabetical heading 'Pigeons'.
42  Anon., *The Dove-cote, or, the Art of Breeding Pigeons*, a Poem, London, 1740, 4.
43  Pehr Kalm, *Kalm's Account of his Visit to England on his Way to America in 1748* (translated by J. Lucas), London, 1892, 300.
    *The Gentleman's Magazine* 24 (first series), 1754, 241.
44  N. Kent, *General View of the Agriculture of the County of Norfolk*, London, 1794, 188.
    J. Boys, *General View of the Agriculture of the County of Kent*, London, 1805, 188.
    H. Holland, *General View of the Agriculture of Cheshire*, London, 1808, 294.
45  Loudon, 1825 (as Note 15), 1048.
46  This is discussed in J. McCann, forthcoming (as Note 28).
47  Waterton, 1839 (as Note 21), 245-6.
48  F. I. Cooke, 'Report on the Farm Prize Competition in Nottinghamshire and Lincolnshire', *Journal of the Royal Agricultural Society* 24 (second series), 1888, 534.
49  Ferguson, 1888 (as Note 3), 415 and *passim*, and Watkins (as Note 3), 20-1
50  These are quoted and examined in McCann, 1991 (as Note 5), 90-116.
51  J. McCann, 'Brick Nogging in the Fifteenth and Sixteenth Centuries, with examples drawn mainly from Essex', *Trans. Ancient Monuments Society* 31, 1987, 108-9.
52  'Excursions', *Proc. Suffolk Institute of Archaeology* 37, part 2, 1990, 171-2, and Suffolk Record Office (Ipswich) HA1/DC3/1 and P638. For this information I am grateful to Leigh Alston and Edward Martin.
53  Colvin and Newman, 1981 (as Note 33), 103.
54  Leigh Alston, 'A Timber Framed Dovecote at Clare', *The Eavesdropper* (Journal of the Suffolk Historic Buildings Group) 4, August 1995.
    E. S. Godfrey, *The Development English Glassmaking*, 1560 - 1640, Oxford, 1975, 55, 204, 209.
55  J. McCann, 'The Origin of Clay Lump in England', *Vernacular Architecture* 28, 1997, 57-67.
56  Daniel Girton, *A Treatise on Domestic Pigeons*, London, 1785, 35.
57  M. C. Heresbach, *Foure Books of Husbandry*, translated and revised by Barnaby Googe, London, 1577, 170.
    Worlidge, J., *Systema Agriculturae*, London, 1669, 192.
58  Messent 1936 (as Note 1), 258. There have been other interpretations of this building - for instance, that it was originally built as a gazebo or garden pavilion. Its present incomplete state makes certainty impossible, but it was described as a dovecote in 1936 and 1947, when it was more complete, and by Mrs. Guild, who has known it all her life.

59 Suffolk Record Office (Ipswich) HB79/2/1. For this information I am grateful to Edward Martin.
60 T. Easton, 'The internal decorative treatment of sixteenth- and seventeenth-century brick in Suffolk', *Post-Medieval Archaeology* 20, 1986, 1-17.
61 N. Pevsner and E. Radcliffe, *The Buildings of England: Suffolk*, 1974, 214.
M. J. A. Beacham, 'Dovecotes in England: an Introduction and Gazetteer', *Trans. Ancient Monuments Society* 34, 1990, 125.
62 McCann, 1997 (as Note 55), 57-8.
63 Messent, 1936 (as Note 1), 258.
64 N. Scarfe, *A Shell Guide to Suffolk*, London, 1976, 72. (This may not be from current observation - perhaps repeated from the first edition of 1960).
Messent, 1936 (as Note 1), 258.
65 Messent, 1936 (as Note 1), 256-7.
66 Pevsner and Radcliffe, 1974 (as Note 61), 352.
67 For this information I am grateful to Edward Martin.
68 G. C. Clopton, *The Clopton Family*, Atlanta, 1984. For this information I am grateful to Edward Martin.
69 Messent, 1936 (as Note 1), 259.
70 For this information I am grateful to Edward Martin.
71 Messent, 1936 (as Note 1), 258.
72 For this information I am grateful to Edward Martin.
73 C. Morley, 'The 1926 Excursions', *Proc. Suffolk Institute of Archaeology* 19, 1926, 234-5.
74 For this information I am grateful to Peter Northeast and David Dymond.
75 J. Webb (ed.), *A Roll of the Household Expenses of Richard de Swinfield, Bishop of Hereford, during part of the years 1289 and 1290*, Camden Society 59 (old series), London, 1854.
76 Cooke, 1920 (as Note 2), 162.
Beacham, 1990 (as Note 61), 125.
77 A. B. Whittingham, *Bury St. Edmunds Abbey*, English Heritage, London, 1971 (based on a paper in *The Archaeological Journal* 108, 1951).
78 Arthur Young, *General View of the Agriculture of the County of Suffolk*, London, 1804, 221.
79 Messent, 1936 (as Note 1), 259.
80 Cooke, 1920 (as Note 2), 188, 274, 279-80, and photographs taken by Frank Pexton.
Messent, 1936 (as Note 1), 259.
Pevsner and Radcliffe, 1974 (as Note 61), 393.
81 For this information I am grateful to Peter Northeast.
82 Priest, 1810 (as Note 23), 39-40, and Loudon, 1825 (as Note 15), 1050.
83 *Cassell's Household Guide*, London, 1870, 39. For this source I am grateful to Beth Davis.
84 J. Munday, 'The Topography of Medieval Eriswell', *Proc. Suffolk Institute of Archaeology* 30, 1966, 201-9. For this source I am grateful to Peter Northeast.
85 For this information I am grateful to Anne Padfield.
86 For this information I am grateful to Mr. H. Russell-Ross and Mr. R. Williams, respectively the owner and architect concerned.
87 Anon., 1735 (as Note 41). The passage was omitted from the 1778 and later editions.
Anon., 1740 (as Note 42), 5.
88 Colvin and Newman, 1981 (as Note 33), 100-1.
89 Mascall, 1581 (as Note 13).
90 Anon., 1735 (as Note 41).
Anon., 1740 (as Note 42), 5.
Colvin and Newman, 1981 (as Note 33), 101.
Girton, 1785 (as Note 56), 32.
91 For example, the Tudor shop at the junction of Lady Street and Water Street, Lavenham, forms an obtuse-angled parallelogram in plan. Every joint of this complex timber-framed building follows the same angle.
92 Anon., 1735 (as Note 41).

93  Mascall, 1581, and Anon., 1735 (as Notes 13 and 41).
94  *The Gentleman's Magazine* 16 (first series), 1746, 478.
95  L. F. Salzman, *Building in England down to 1540*, Oxford, 1952, 185.
    Ferguson, 1888 (as Note 3), 418-9.
96  Godfrey, 1975 (as Note 54).
97  McCann, 1991 (as Note 5), 120.
98  J. Whitaker, *A Descriptive List of the Medieval Dovecotes in Nottinghamshire*, Mansfield, 1927, 58, 61, 100-1.
99  McCann, 1991 (as Note 5), 117-20.
100 Elihu Burritt, *Walks in the Black Country, and its Green Borderland*, Kineton, 1976, 121-3.
101 Davis, 1986 (as Note 1), 69-71, 77, 79-81.
102 Pipes are present in the great Tudor dovecote at Willington, Bedfordshire (National Trust), and at Hedingham Castle, Essex (J. McCann, *Essex Archaeology and History* 28, 1997, 294-8). Another in Essex is illustrated in McCann, 1991 (as Note 5), 128.
103 Colvin and Newman, 1981 (as Note 33), 102.
104 Priest, 1810 (as Note 23), 39.
105 Beazley, 1966 (as Note 36), 107.
106 The passages are quoted in full in McCann, 1997 (as Note 55), 57-8.
107 D. Bouwens, 'Clay Lump Windmills', *Newsletter of the East Anglian Earth Buildings Group* 4, April 1997, 6-7.
    C. J. W. Messent, *The Old Cottages and Farmhouses of Norfolk*, Norwich, 1928, 190, 209.
108 McCann, 1997 (as Note 55).
109 Cooke, 1920 (as Note 2), 22-3, 88-9.
110 D. C. Bailey and M. C. Tindall, 'Dovecotes of East Lothian', *Trans. Ancient Monuments Society* 6 (new series), 1963, 27-30.
111 Colvin and Newman, 1981 (as Note 33), 103.
112 An easily accessible example is the dovecote from Haselour Hall, Staffordshire, which has been rebuilt at the Avoncroft Museum of Buildings, Bromsgrove, Worcestershire. The sixteenth-century timber frame survives within an eighteenth-century brick envelope.
113 Davis, 1986 (as Note 1), 70 and *passim*.
114 Young, 1804 (as Note 78), 221.
115 See Chapter 10.
116 Kent, 1794 (as Note 44), 188.
117 Rudge, 1807 (as Note 16), 326.
    Priest, 1810 (as Note 23), 332.
118 W. and H. Raynbird, *On the Agriculture of Suffolk*, London, 1849, 48-51, 115-8, 124-5, 131-2.
119 Major improvements to the house in the early eighteenth century are discussed in 'Excursions', *Proc. Suffolk Institute of Archaeology* 35, part 3, 1995, 365-6.
120 For this information I am grateful to Edward Martin.
121 Pevsner and Radcliffe, 1974 (as Note 61), 197, 375, 412, 437.
122 The National Monuments Record Centre, Swindon (Henham parish file).
    A. Mackley, 'The Construction of Henham Hall', *Georgian Group Journal* 6, 1996.
123 On Benacre, personal communication from Michael Seago, 13 December 1992.
124 I am grateful to Timothy Easton for introducing me to this building.
125 Personal communication from Mark Barnard, 8 September 1994.
126 Suffolk Record Office (Bury) FL 667/5/12.
127 J. McCann, 'The Conservation of Historic Dovecotes', *Journal of Architectural Conservation* 1, no. 2, July 1995, 78-96.
128 The East Anglian Earth Buildings Group (known as EARTHA) publishes Newsletters and organises demonstrations of the techniques used to repair historic buildings of earth materials. It can be contacted through any County or District Council in East Anglia.

# Appendix
## by Dr. Rosemary Hoppitt

*Suffolk dovecotes: some evidence of medieval dovecotes from manorial account rolls*

The list in this appendix includes references to dovecotes (*columbare*) in (mainly) thirteenth- and fourteenth-century manorial account rolls for various Suffolk manors. It is not complete, consisting only of those manors with documents deposited at the Suffolk Record Office (Ipswich branch). As such it represents work in progress, and should not be considered comprehensive. The intention is to record the existence of dovecotes across Suffolk during this period, and ultimately to map a county distribution.

Dovecotes are mentioned in account rolls for two main reasons. They appear under expenses when they are built or repaired, and they appear under income when the pigeons are being sold or the dovecote rented out. The expenses are useful in gaining clues about the structure and appearance of medieval dovecotes, identifying as they do the craftsmen and labourers working on them as well as the materials being used. Incomes vary considerably, with some manors receiving little or no income from the dovecote by way of sales (which does not necessarily mean that pigeons were not being produced for the manorial table), and others with sales in hundreds of pigeons.

The list below includes the manor named in the account roll, and the reference number of the document (all are at Suffolk Record Office, Ipswich, unless otherwise indicated). In some the reference is to a collection of a number of account rolls, in which case the years quoted are only those for which details are given. The detail provided is selective, and designed to give a flavour of the type of references made in the document, which give a pointer to the appearance of and levels of production from these Suffolk examples.

| *Manor and year* | | *Detail and source* |
|---|---|---|
| Akenham, | 1387/9 | dovecote (HD 1469/7) |
| | 1392/3 | income from the rent of the dovecote 6/8d (HD 1469/8) |
| Easton Bavents, | 1330/1 | one man paid for four days for roofing the dovecote (V5/19/1.1-1.5) |
| | 1356/7 | no income from the dovecote because it is held by the steward of the hospitium |
| | 1382/3 | four hundred laths purchased for the bakehouse and the dovecote 16d ... iron nails bought for the same 8d |
| | 1394/5 | three dozen and three (39) pigeons |
| | 1437 | 3/9d received from (the sale of) 135 pigeons being the total from the dovecote there this year ... pigeons three for a penny |
| Foxhall (Casnells), | 1283 | income from the dovecote 3/4d (HA 50/3/94) |
| Framlingham, | 1277/8 | from the dovecote nothing this year because it is broken down (HD 1538/225/1) |
| | 1385/6 | 431 pigeons received this year from the dovecote within the great pond below the castle at Framlingham ... and 139 pigeons received as a gift from Matille Atiltone (Hospice Account)[1] |

# The Dovecotes of Suffolk

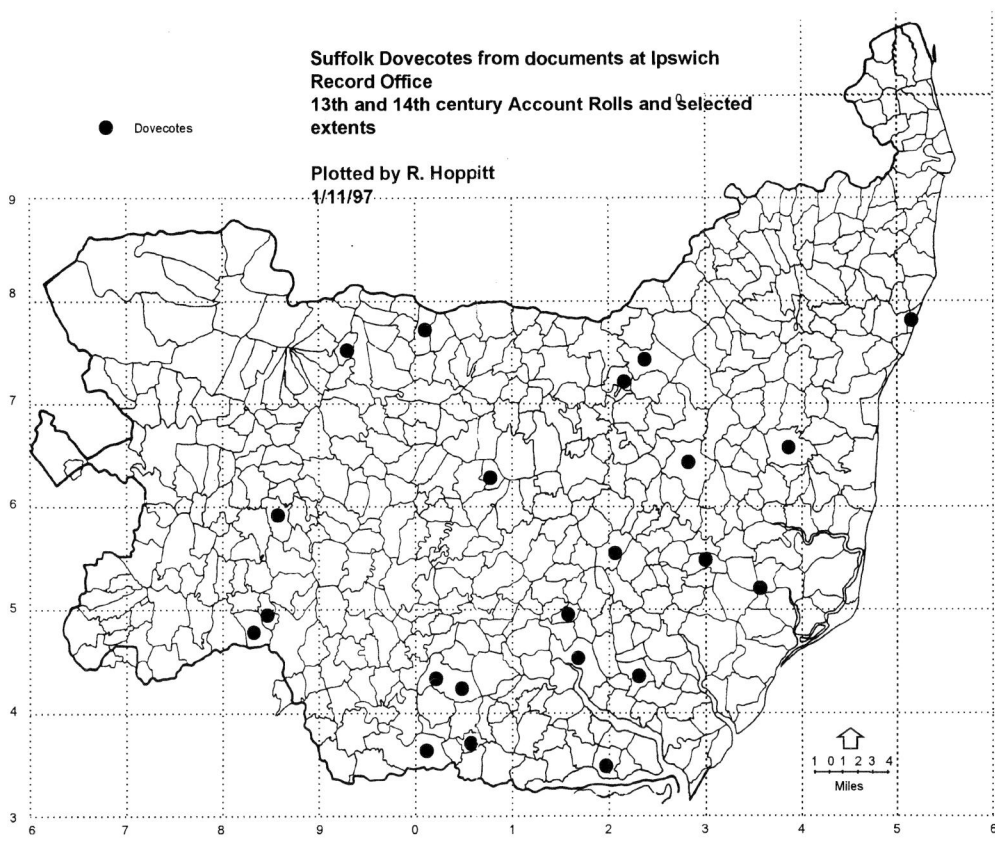

| | | |
|---|---|---|
| Gipping, | 1401 | 10/2½d for 307 pigeons sold at three a penny sum 10/2½d. There remains at the time of the account 111 pigeons ... of which in tithe 56 pigeons and allowed to William Slueut (?) by the lord for his expenses 164 pigeons (HD 1538/236/15-23) |
| | 1403 | 2/4d from four-score and eleven (91) pigeons sold at three for a penny |
| Glemsford, | 1330 | dovecote (HD 64/1/2) |
| Hadleigh, | 1305 | dovecote valued at 3/- (PSIA 3, Extent of Hadleigh Manor)[2] |
| Hadleigh (Pond Hall), | c. 1400 | dovecote (GC 16/1/1 Extent of Pond Hall manor) |
| Harkstead, | 1320/2 | (payment to) a cooper ... plastering the dovecote ... for wood and wattle for mending the dovecote (S1/10/6.6) |
| Hawstead, 14th century extent | | dovecote (HA 246/C6/1) |
| Holton St. Mary, | 1345 | 40d from the profits of the dovecote this year (HA 246/A4/7) |
| Horham, | 1343 | 12d from the dovecote this year (HA 85/484/136) |
| Ipswich (Holy Trinity), | 1351/2 | (payment to) one man ... for mending and roofing the grange and dovecote and the walls of the dovecote (HD 1538/271/4-7) |
| Kelsale, | 1277/8 | costs of the dovecote ... making 1000 shingles and constructing the dovecote 9/- 1000 nails for the same 5/- sum 14/- (HD 1538/8) |

| | | |
|---|---|---|
| Loudham (Pettistree), | 1379 | (payment to) carpenter ... for building the dovecote (HD 1538/295/3) |
| Otley, | 1344 | dovecote (HB 11/1/136) |
| Sapiston, | 1317 | one lock bought for the dovecote 1½d ... building of dovecote ... ridging the roof 21d (HD 1538/338/15-24) |
| | 1337 | new roofing and ridging of the lord's dovecote ... whitening the dovecote 5d ... lime bought 1d (HD 1538/338/25-34) |
| | 1340 | one lock and key bought for the dovecote 3d |
| | 1345/6 | from the dovecote during the time of this account 411 pigeons |
| | 1350 | from the dovecote 680 pigeons this year |
| Stanstead, | 1338 | from the dovecote nothing this year because it is ruinous and broken down and in store (?)(HA 30/50/22/20-27) |
| Stoke by Nayland, | 1386/7 | from the dovecote 2/1d received from the sale of 62 pigeons this year five for 2d (HA 246/A8/13) |
| Stradbroke (Barlow Hall), | 1350 | from the dovecote 57 pigeons (HD 1070/17-19) |
| | 1352 | wages of one roofer (cooper) for 1 day roofing over the dovecote in places 4d |
| Thelnetham, | 1375 | from the dovecote nothing this year due to lack of repair (HD 1538/380/1-8) |
| | 1385 | this account roll contains detail of the construction of the dovecote including the men employed to prepare and construct with wattle and daub and to dig clay for the daub; a carpenter was employed preparing pins (*pynnes*) and making floors of boards (*flores de bord*); underwood was purchased for the construction work. In all the costs amounted to 45/1½d (HD 1538/380/9-13) |
| Wantisden, | 1398 (?) | 3/4d from the farm of the dovecote |
| | | (payment to) one roofer (cooper) for 12 days roofing over the grange and the dovecote ... 4/- at 4d per day (HD1538/406/23-24) |

## NOTES:

1   J. Ridgard (ed.), *Medieval Framlingham: Select Documents 1270-1524*, Suffolk Records Society, Woodbridge, 1985, 118-9.
2   Hugh Pigot: 'Hadleigh the Town, the Church and the Great Men': Appendix A Hadleigh Extent', *Proc. Suffolk Institute of Archaeology* 3, 1863, 229-52.